Victor Gutenmacher
N.B. Vasilyev

Lines and Curves

A Practical Geometry Handbook

Based on an English translation of the original
Russian edition by A. Kundu

Foreword by Mark Saul

Birkhäuser
Boston • Basel • Berlin

Victor Gutenmacher   N.B. Vasilyev (deceased)
21 Westbourne Terrace, #4
Brookline, MA 02446
U.S.A.

Original translation by
Anjan Kundu
Saha Institute of Nuclear Physics
Theory Group
Calcutta 700 064
India

Cover illustrations by Michael Panov.

AMS Subject Classifications: Primary: 51-XX, 51M05, 51Nxx; Secondary: 00A69, 65D18, 65D17, 51-01

Library of Congress Cataloging-in-Publication Data
Gutenmakher, V. L. (Viktor L'vovich)
    [Priamye i krivye. English]
    Lines and curves : a practical geometry handbook / Victor Gutenmacher, N.B. Vasilyev.
        p. cm.
    "Based on an English translation of the Russian edition by A. Kundu."
    Vasilyev's name appears first on the earlier edition.
    Includes index.
    ISBN 0-8176-4161-0 (acid-free paper)
        1. Geometry–Problems, exercises, etc. I. Vasil'ev, N. B. (Nikolai Borisovich) II.
Vasil'ev, N. B. (Nikolai Borisovich) Straight lines and curves. III. Title.

QA459.V3613 2004
516'.0076–dc22   2004051906

ISBN 0-8176-4161-0   Printed on acid-free paper.

©2004 Birkhäuser Boston, Inc.   *Birkhäuser*   ®
Based on an English translation of the original Russian edition by A. Kundu, *Straight Lines and Curves*, Mir Publishers, Moscow, ©1980.

Printed in the United States of America.   (TXQ/MP)

9 8 7 6 5 4 3 2 1   SPIN 10746420

Birkhäuser is a part of *Springer Science+Business Media*

www.birkhauser.com

Foreword

Tolstoy begins his epic novel by cutting to the heart of the human experience. Starting with a picture of a fashionable Moscow soirée, he shows how the events of the day, which are about to overwhelm his characters, first begin to encroach on their consciousness.

In just this way, the authors of *Lines and Curves* plunge their reader directly into the mathematical experience. The do not start with a set of axioms, or a list of definitions, or with an account of the concept of a locus. Rather, they present a simple, matter-of-fact discussion of a cat sitting on a moving ladder. This leads to various problems about the trajectories of moving points. The reader is led, almost imperceptibly, to classic results in synthetic geometry, looked at from a new perspective (p. 11), then on to new problems, whose traditional synthetic solutions are considerably more complicated than those offered here (p. 13), and finally through a variety of problems whose difficulty gradually increases as we work them.

Things begin to look more traditional beginning with Chapter 2, as the authors provide a toolkit (an "Alphabet") for solving new problems, along with ways to combine these tools ("Logical Combinations"), and a new context ("Maximum and Minimum") in which to apply them. The structures of mathematics unfold naturally, and do not overwhelm the reader as they begin to encroach on his or her consciousness.

This is not to say that the material is simple—only that it is rendered simple by the authors' art. There are difficult problems here. There are sometimes messy calculations and intractable formulas. But in every case, a careful analysis, informed by the results of previous pages, will lead to further insight, to more general conclusions, and ultimately to the solution of more sophisticated problems.

Slowly and naturally, the formality of mathematics emerges in a way which demonstrates its utility, not just its complexity or its efficiency. Indeed, the artful exposition makes the formalization seem the most intuitive way to express the result discussed, the informal analysis only a necessary step towards this deeper understanding. The discussion of level curves is an example of this: the "alphabet" of the early chapters is revisited in the language of functions defined on the plane. The reader is caught thinking, "I knew that! But now I know how to say it."

The authors' art reaches a wide range of readers. The novice will be gently charmed by the cat and the tiny rings, and led through material from a standard mathematics course—analytic definitions of conics, level curves and functions, advanced geometry of the triangle. The experienced reader will revisit old friends from new perspectives, as the authors reveal a new way of thinking about classic results. The expert will find new relationships among old friends, and new ways to think about classic results.

And the same reader may enjoy each of these experiences. For this is a book to be read more than once, to be savored differently at different times in one's development. I hope you, dear reader, will enjoy your acquaintance with it as much as I have, whether this be your first reading or your nth.

Mark Saul
Bronxville High School (ret.)

Gateway Institute
City University of New York

Preface

Definitions are very important. In all of mathematics, definitions tell us where to begin, and the right definitions make it easy for us to move forward from there. Careful definitions can often turn difficult concepts into clear and understandable ones, and they can make our intuition precise. The intuitive notion of a circle, for instance, is made precise when we define it as the set of points equidistant from a fixed point (the center); thus a definition can transform figures into mathematical language. For some people, a good definition is even a work of art. The great mathematician Alexander Grothendieck once wrote, "Around the age of twelve... I learned the definition of a circle. It impressed me by its simplicity and its evident truth, whereas previously, the property of 'perfect rotundity' of the circle had seemed to me a reality mysterious beyond words. It was at that moment... that I glimpsed for the first time the creative power of a 'good' mathematical definition... still, even today, it seems that the fascination this power exercises over me has lost nothing of its force."

Grothendieck's reaction underscores how definitions can simplify "mysterious" mathematical ideas. As we will see, the same mathematical term may be defined in different ways. For example, a circle can also be described as an algebraic equation or as the trajectory of a moving point: Both these interpretations appear in our first problem in the Introduction, which is about the path of a cat that sits on a falling ladder. Throughout the book, different definitions for the same geometric object are extremely useful, and the choice of an appropriate definition is a key to solving many of the questions in this text, from the first section to the last.

Lines and Curves is divided into chapters that contain sets of problems; each set is written like a short lecture, and the example problems are accompanied by complete solutions. We emphasize the geometric properties of paths traced by moving points, the loci of points that satisfy given geometric constraints, and problems of finding maxima and minima. Again and again, we will view geometric shapes not as static figures in space, but as points and lines *in motion*: we want to encourage the reader to look at geometry in this new light, to reformulate problems in the context of rotating lines, moving circles, and trajectories of points. *The language of motion, in fact, allows us to construct intuitive, straightforward proofs for concepts that can otherwise be very complicated.* Overall, there are more than 200 problems that lead the reader from geometry into important areas of modern mathematics. Some of these problems are elementary, others quite involved, but there is something for everyone. Hints and solutions to many of the exercises are also provided in an appendix.

The book can be used with an interactive educational software package, such as The Geometer's Sketchpad®, for exploring the loci of points and constructions. Not only can students work through the solutions, they can simultaneously draw diagrams using a pencil and paper or using various computer tools. Indeed, the book is meant to be read with pencil and paper in hand; *readers must draw their own diagrams in addition to following the figures we have given.* We hope that our readers will participate in each of our investigations by drawing diagrams, formulating hypotheses, and arriving at the answers—in short, by joining us in our experimental approach. For this edition, we have included various constructions for the students to recreate alongside each section; these are repeated in the newly added final chapter on drawings and animation, and they demonstrate the importance of the experimental aspect in the study of geometry.

We use a few special symbols throughout. Our main character, the Cat, reappears in many of the chapters and is an emblem for the book as a whole. (Cats, naturally, are perfect geometers—consider Problem 4.10, which involves finding an optimal position: cats solve such real-life problems every day.) Vertical lines on the left indicate the statements of the problems. Depending on its position, the question mark (?) symbolizes the words "exercise," "verify," "think about why this is true," "is this clear to you," etc. Problems are flanked at beginning and end by a small white square, and an arrow (↓) after a problem statement indicates that the solution is provided in the Appendix "Answers, Hints, and Solutions" at end of the book. Particularly challenging problems are distinguished by an asterisk (∗) next to the problem number. *We assume that the reader is already well-versed in the fundamentals of Euclidean geometry*, but for reference, we include a list of useful geometric facts and formulas in Appendices A and B. Finally, Appendix C, "A Dozen Assigments," features additional exercises which serve to clarify and extend the theorems and concepts presented in the main text.

While *Lines and Curves* originated as a geometry textbook for high school students in the I. M. Gelfand Multidisciplinary Correspondence School, its prerequisites include basic plane and analytic geometry. Furthermore, its wide range of problems and unique kinematics-based approach make *Lines and Curves* especially valuable as a supplemental text for undergraduate courses in geometry and classical mechanics.

This newly revised and expanded English edition is dedicated to my coauthor Nikolay Vasilyev, a great friend and longtime collaborator. Nikolay died in 1998. It was a joy to write this book with him.

We are deeply grateful to Dr. I. M. Gelfand, whose advice influenced the first Russian edition. We would also like to thank I. M. Yaglom, V. G. Boltyansky, and J. M. Rabbot, who read the initial manuscript; and T. I. Kuznetsova, M. V. Koleychuk, and V. B. Yankilevsky, who illustrated previous editions.

Many remarkable people helped with this new edition: Joseph Rabbot, Wally Feurzeig, Paul Sontag, Garry Litvin, Margaret Litvin, Anjan Kundu, Yuriy Ionin, Tanya Ionin, Elizabeth Liebson, Victor Steinbok, Sergey Bratus, Ilya Baran,

Michael Panov, Eugenia Sobolev, Sanjeev Chauhan, Olga Itkin, Olga Moska, Lena Moskovich, Philip Lewis, Yuriy Dudko, and Pierre Lochak. It was Pierre Lochak who provided the English translation (from the original French) for the Grothendieck quote. Not only am I thankful to each of these individuals for their help, I am also happy to call them friends.

Finally, I wish to thank the entire staff at Birkhäuser and the anonymous reviewers whose input significantly improved the text. In particular, I am grateful to Avanti Athreya, who painstakingly edited the book and gave me wonderful suggestions from beginning to end. I am also indebted to Elizabeth Loew and John Spiegelman for their exceptional efforts in production, and to Tom Grasso and Ann Kostant for their unflagging support and attention to detail. Without their collective assistance, this work would not have seen the light of day—at least in its current form.

Victor Gutenmacher
Boston, MA

Contents

Notation

$|AB| = \rho(A, B)$ the length of the segment AB
(the distance between the points A and B).

$\rho(A, l)$ the distance from the point A to
the straight line l.

$\widehat{ABC}$ the value of the angle ABC
(in degrees or radians)

$\overset{\frown}{AB}$ the arc of a circle with endpoints A and B.

$\triangle ABC$ triangle ABC

S_{ABC} the area of the triangle ABC

$\{M : f(M) = c\}$ the set of points M, which
satisfies the condition $f(M) = c$

Lines and Curves

Introduction

Introductory problems

0.1. A ladder standing on a smooth floor against a wall slides down onto the floor. Along what curve does a cat sitting in the middle of the ladder move?

Let us suppose our cat is calm and sits quietly. Then we can see that behind this picturesque formulation is the following mathematical problem.

A right angle is given. Find the midpoints of all the possible segments of a given length d whose endpoints lie on the sides of the given angle.

Let us try to guess what sort of set this is. Obviously, as the line segment rotates with its endpoints sliding along the sides of the angle, its center describes a certain portion of a curve. (This is obvious from the first statement of the problem.) First, let us determine where the endpoints of this curve lie. They correspond to the extreme positions of the segment when it is vertical or horizontal. This means that the endpoints A and B of the line lie on the sides of the angle at a distance $d/2$ from its vertex.

Let us plot a few intermediate points of this curve. If you do this accurately enough, you will see that all of them lie at the same distance from the vertex O of the given angle. Thus, we can say that:

The unknown curve is an arc of a circle of radius $d/2$ with center at O. Now we must prove this.

☐ We shall first prove that the midpoint M of the given segment KL (where $|KL| = d$) always lies at a distance $d/2$ from the point O. This follows from the fact that the length of the median OM of the right-angled triangle KOL is equal to half the length of the hypotenuse KL. (One can easily be convinced of the validity of this fact by extending the triangle KOL to form the rectangle $KOLT$ and by recalling that the diagonals KL and OT of the rectangle are equal in length and are bisected by the point of intersection M.)

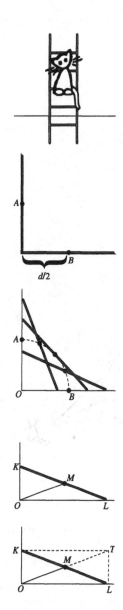

1

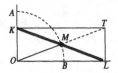

Thus, we have proved that the midpoint of the segment KL always lies on the arc $\overset{\frown}{AB}$ of a circle with center O. This arc is the set of points we were looking for.

Strictly speaking, we have to prove also that an arbitrary point M of the arc $\overset{\frown}{AB}$ belongs to the unknown set. It is easy to do this. Through any point M of the arc $\overset{\frown}{AB}$ we may draw a ray OM, mark off the segment $|MT| = |OM|$ along it, drop perpendiculars TL and TK from the point T to the sides of the angle and the required segment KL with its midpoint at M is constructed. □

The second half of the proof might appear to be unnecessary: It is quite clear that the midpoint of the segment KL describes a "continuous line" with endpoints A and B; this means that the point M passes through all of the arc $\overset{\frown}{AB}$ and not just through parts of it. This analysis is perfectly convincing, but it is not easy to put it into strict mathematical form.

Let us now consider the motion of the ladder (from Problem **0.1**) from another point of view. Suppose that the segment KL (the "ladder") is fixed and the straight lines KO and LO ("the wall" and "the floor") rotate correspondingly about the points K and L so that the angle between them is always a right angle. The fact that the distance from the center of the segment to the vertex O of the right angle always remains the same can, in fact, be reduced to a well-known theorem: *If two points K and L are given in a plane, then the set of points O for which the angle $\overset{\frown}{KOL}$ equals $90°$ is a circle with diameter KL.* This theorem and its generalization, which will be given in Proposition **E** of Chapter 2, p. 22, will frequently help us in the solution of problems. Let us return to Problem **0.1** and ask a more general question.

0.2. Along what curve does the cat move if it does not sit in the middle of the ladder?

In the figure a few points on one such curve are plotted. It can be seen that it is neither a straight line nor a circle; i.e., it is a new curve for us. The method of coordinates—that is, the principles of ana-

2

lytic geometry—can be used to help us determine out what sort of curve this is.

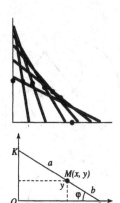

☐ We introduce a coordinate system by regarding the sides of the angle as the x and y-axes. Suppose the cat sits at some point M (x, y) at a distance a from the endpoint K of the ladder and at a distance b from L (where $a + b = d$). We shall find the equation connecting the x and y coordinates of the point M.

If the segment KL is inclined to the axis Ox at an angle φ, then $y/b = \sin \varphi$ and $x/a = \cos \varphi$; hence, for any arbitrary φ $\left(0 \leq \varphi \leq \frac{\pi}{2}\right)$

$$\frac{x^2}{a^2} + \frac{y^2}{b^2} = 1. \qquad (1)$$

The set of points whose coordinates satisfy this equation is an *ellipse*. Hence the cat will move along an ellipse. ☐

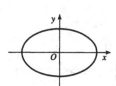

Note that when $a = b = d/2$, the cat sits in the middle of the ladder, and equation (1) becomes the equation of a circle $x^2 + y^2 = (d/2)^2$. We thus get one more solution—this time, an analytical solution—to Problem **0.1**.

The result of Problem **0.2** explains the construction of a mechanism for drawing ellipses. This mechanism, which is shown in the figure, is called *Leonardo da Vinci's ellipsograph*.

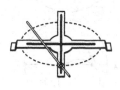

Copernicus' Theorem

0.3. Inside a stationary circle is another circle whose diameter is half the diameter of the first circle; this smaller circle touches the larger one from within, and it rolls along the larger circle without sliding. What line does the point K of the moving circle describe?

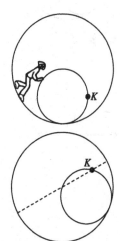

The answer to the problem is astonishingly simple: the point K moves along a *straight* line—more correctly, along a diameter of the stationary circle. This result is called *Copernicus' Theorem*.

Try to convince yourself of the validity of this theorem by experiment. (It is important here that the inner circle rolls without sliding, i.e., the lengths of the arcs rolling against each other are equal.) It is not difficult

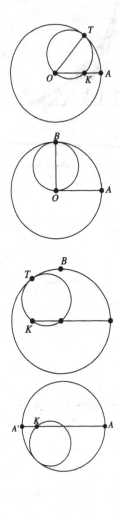

to prove; we need only recall the theorem about the inscribed angle.

☐ Suppose that the point of the moving circle, which occupies position A on the stationary circle at the initial instant, has come to the position K, and that T is the point of contact of the circle at the present moment of time. Since the lengths of the arcs $\overset{\frown}{KT}$ and $\overset{\frown}{AT}$ are equal and the radius of the movable circle is half as large as that of the fixed circle, the angular size of the arc $\overset{\frown}{KT}$ in degrees is double that of the arc $\overset{\frown}{AT}$. Therefore, if O is the center of the stationary circle, then according to the *theorem on the inscribed angle* (see p. 11), $\overset{\frown}{AOT} = \overset{\frown}{KOT}$. Hence, the point K lies on the radius AO.

This argument holds until the moving circle has rolled around one quarter of the bigger circle (the circles then touch at the point B of the bigger circle, for which $\overset{\frown}{BOA} = 90°$ and K coincides with O). After this, the motion will be continued in exactly the same way—the whole picture will simply be reflected symmetrically about the straight line BO and then, after the point K reaches the opposite end A' of the diameter AA', the circle will roll along the lower half of the stationary circle and the point K will return to A. ☐

Let us compare the results of problems **0.1** and **0.3**. They are attractive for the following reason. Both problems deal with the motion of figures (the first with the motion of a segment, the second with the motion of a circle). The motion is quite complicated, but the paths of certain points appear to be unexpectedly simple. These two problems turn out to be not only related in appearance, but even the motions themselves, as discussed in in the problems, coincide with each other.

Indeed, suppose a circle of radius $d/2$ rolls along the inside of another circle of radius d, and suppose KL is the diameter of the moving circle rigidly fixed to it. According to Copernicus' Theorem, the points K and L move along stationary straight lines (along the diameters AA' and BB' of the bigger circle, respectively). Thus, the diameter KL slides with its endpoints along

4

two mutually perpendicular straight lines, i.e., it moves just like the segment in the Problem **0.1**.

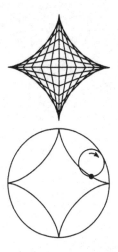

We conclude with one more interesting problem connected with the motion of the segment KL: What set of points is covered by this segment–that is, what is the union of all possible positions of the segment KL during its motion? The curve bounding this set is called an *astroid*. It is possible to construct this curve in the following way: Let a circle of diameter $d/2$ roll inside another circle of diameter $2d$, and draw the trajectory of any particular point of the rolling circle. This trajectory will be an astroid. We shall discuss this curve and its close relatives in Chapter 7 of our book, where the reader will make a more detailed acquaintance with the interconnections among the problems we have discussed.

However, before discussing such intricate problems and curves, let us pay careful attention to problems dealing specifically with straight lines and circles. Other types of curves will not appear in the first five chapters.

CHAPTER 1

Sets of Points

In this chapter we discuss the basic statements and problems in this book, and we illustrate them with a number of examples. We also provide an arsenal of concepts and methods for solving certain geometry problems. The chapter ends with a set of various geometric exercises.

We first discuss the most-frequently used term in the book. Indeed, it is in the title of this very chapter: a "*set of points.*"

The concept of a "*set of points*" is very general. A set of points could be any figure, one point or several, a line or a domain in a plane.

In many of the problems here we have to find a set (or locus) of points that satisfies a certain condition. Answers to such problems are, as a rule, figures known from school geometry (straight lines, circles, sometimes pieces into which these lines divide a plane, etc.). The main task is to guess what sort of a figure the answer is. Thus, in Problem **0.1** about the cat, we guessed the answer—it was a circle, and in Problem **0.3** the answer turned out to be a straight line.

To completely solve these problems, however, we have to carry out a more thorough examination. We must establish the following:

(a) *All the points satisfying the given condition belong to the figure;*

(b) *All the points of the figure satisfy the given condition.*

Sometimes both statements are obvious, the direct statement as well as its converse; sometimes only one

7

of them is obvious. Sometimes it is even difficult to guess the answer.

Let us consider a few typical problems.

1.1. A point O lies on a segment AC. Find the set of points M for which $\widehat{MOC} = 2\widehat{MAC}$.

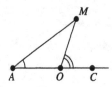

□ *Answer*: The union of the circle with center O and radius $|AO|$ (omitting the point A) and the ray OC (omitting the point O).

Let us verify this claim. Suppose the point M of the unknown set does not belong to the straight line AO. We shall prove that the distance $|MO|$ from the point M to the point O is equal to $|AO|$. Let us construct the triangle OAM. According to the theorem on the exterior angle of a triangle, the angle MOC is equal in magnitude to the sum of the two interior angles not adjacent to it at A and M:

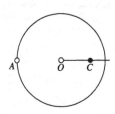

$$\widehat{OAM} + \widehat{AMO} = \widehat{MOC} = 2\widehat{MAO}.$$

From the condition of the problem it follows immediately that $\widehat{OAM} = \widehat{AMO}$. Hence, AMO is an isosceles triangle, i.e., $|OM| = |AO|$. Therefore, M is on the circle we described.

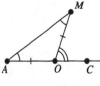

We shall now prove the validity of the converse statement: namely, that the condition is satisfied by any point M on the circle described in the answer.

The triangle AMO is evidently isosceles; the values of its angles A and M are equal, and by the same theorem concerning the exterior angle, $\widehat{MOC} = 2\widehat{MAC}$.

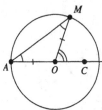

Suppose now the point M belongs to the ray OC, $M \neq O$. Then $\widehat{MOC} = 2\widehat{MAC} = 0$, and the condition is satisfied.

The remaining points on the straight line AO do not belong to the unknown set. For such points M, the angles MOC and MAC are either both 180 degrees in measure, or one of the angles is 180 degrees and the other is zero (about the point O, however, one can say nothing). □

1.2. Two wheels of radii r_1 and r_2 $(r_1 > r_2)$ roll along a straight line l. Find the set of points of intersection M of their interior common tangents (see the figure).

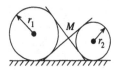

8

□ *Answer*: A straight line, parallel to l.

Note that the point M lies on the axis of symmetry of the two circles, i.e., on the straight line O_1O_2, where O_1 and O_2 are the centers of the circles. Therefore, one can look for the set of points of intersection of the straight line O_1O_2 with one of the tangents T_1T_2.

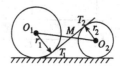

Let us consider two such circles and let us draw their radii O_1T_1 and O_2T_2 to the points of tangency. We see that the point M divides the segment O_1O_2 in the ratio r_1/r_2 (the right-angled triangles MO_1T_1 and MO_2T_2 are similar). It is clear that the set of centers O_1 and the set of centers O_2 are straight lines parallel to the straight line l. The set of points M which divide the segments O_1O_2, with endpoints on these straight lines, in the fixed proportion r_1/r_2, is itself a straight line parallel to l.

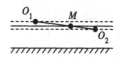

Thus, the set of points of intersection of the tangents is a straight line parallel to the line l and placed at a distance $2r_1r_2/(r_1 + r_2)$ from this line ⟨?⟩. Note that the conclusion is independent of O_1 and O_2. □

The next problem demands a more careful examination. We have to divide the plane into several parts and carry out a separate argument for each of those parts.

1.3. Given a rectangle $ABCD$, find all points in the plane such that the sum of the distances from each point to two straight lines AB and CD is equal to the sum of the distances to the straight lines BC and AD.

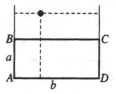

□ Let us denote the lengths of the sides of the rectangle by a and b. First, consider a rectangle which is not a square: let $a < b$.

The points lying inside the rectangle and also between the extensions of its larger sides do not satisfy the requirements of the problem, since one sum of the distances is equal to a and the other is not less than b.

Let the point M now lie between the extensions of the smaller sides of the rectangle. Let y denote the distance from M to the nearest of the longer sides of the rectangle. Then its distance from the opposite side is equal to $y+a$. For the point to satisfy the requirement

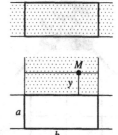

9

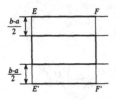

of the problem, the equality $y + (y + a) = b$ must hold, from which it follows that $y = (b - a)/2$. Therefore, among the points located between the extensions of the smaller sides of the rectangle, those and only those which lie at a distance $(b - a)/2$ from the closer and larger sides of the rectangle satisfy the condition. The set of points in this domain is the union of two segments EF and $E'F'$.

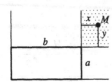

Finally, we shall consider an arbitrary point M lying in the angle between the extensions of the two neighboring sides BC and DC of the rectangle. Let us denote by x and y the distances from the point M to the straight lines CD and BC, respectively. Then one can express the requirement of the problem as $x + (x + b) = y + (y + a)$ or $y = x + (b - a)/2$.

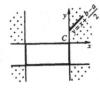

Note that the numbers x and y can be regarded as coordinates of the point M in the coordinate system with the axes Cx and Cy. In this coordinate system the equation $y = x + (b - a)/2$ defines a straight line parallel to the bisector of the angle xCy. Thus we have proved that among the points lying in the angle under consideration, those and only those which lie on the straight line $y = x + (b - a)/2$ satisfy the requirement of the problem.

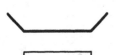

We can use the same argument for the remaining three angles. We have thus analyzed all the points of the plane. The set of all the points that satisfy the stated requirement is plotted in the figure.

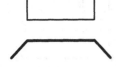

We also have to consider the case when the rectangle is a *square*, i.e., when $a = b$, and to determine what set the required set of points reduces to.

It can easily be seen that it will be the union of the square and the extensions of its diagonals ⟨?⟩. □

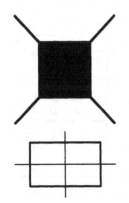

Note that *since a rectangle has two axes of symmetry and the pairs of its symmetrical sides in the given conditions are totally identical, the unknown set of points must also have those two axes of symmetry.* Therefore, in the solution it was sufficient to only consider any one of the quadrants into which the plane is divided by these axes, and not the whole plane.

In the case of a square, all four axes of symmetry of the square are also axes of symmetry of the set we are looking for.

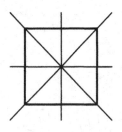

A family of lines and motion

Together with sets of *points* we shall also consider sets of lines or, as they are frequently called, *families of lines*.

In geometry problems that involve a family of circles or straight lines, *it is often convenient to imagine the family as a moving circle or a straight line*. We have already formulated and solved our first few problems in the language of motion, and we shall use this language repeatedly in what follows. Indeed, many problems and theorems can be explained more vividly if we reformulate them in the context of points and lines in motion.

We don't have to look far for an example. Let us return to Problem **1.1**. The result we found there can be given as follows:

Suppose the straight line AM rotates about the point A with constant angular velocity ω (i.e., it turns through an angle of magnitude ω in unit time) and the straight line OM rotates abut the point O with angular velocity 2ω; at the initial point of time both lines coincide with the straight line AO. Then the point of intersection M of straight lines moves along a circle with center O.

From this we can obtain a well-known *theorem on the inscribed angle*. If in time t the straight line AM rotates from the position AM_1 to the position AM_2 through an angle ϕ, then the straight line OM rotates through an angle 2ϕ or, in other words, the *magnitude of the inscribed angle M_1AM_2 is half the magnitude of the corresponding central angle M_1OM_2.*

One can formulate this theorem more vividly as follows.

A theorem about a tiny ring on a circle. A small ring is put on a wire circle. A rod which passes through this ring rotates around the point A of the circle. *If the rod rotates uniformly with an angular velocity ω, the ring in this case moves uniformly around the circle with an angular velocity 2ω.*

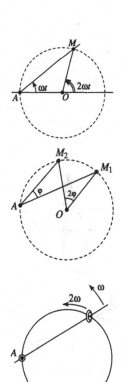

11

Let us give one more example of a theorem which may be formulated in the language of motion.

Suppose the straight line *l* describes a *uniform translation* in a plane, i.e., it moves in such a way that its direction remains unchanged and its point of intersection *M* with a certain stationary straight line *m* moves uniformly along the line *m*. Then, *the point of intersection N of the line l with any other stationary straight line n also moves uniformly.* This is, in fact, a restatement of the theorem which states that *parallel straight lines cut off proportional segments on the sides of an angle.* To make an analogy with the theorem about the ring on a circle, we can express this in the following way.

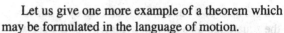

A theorem about a tiny ring on a straight line. A small ring is placed at the point of intersection of two straight lines. *If one of the lines is fixed and the other describes a uniform translation* (parallel to itself), *the ring also moves uniformly.* We shall encounter various families of straight lines later on.

When one has to deal with a family of straight lines passing through a single point or parallel to a fixed direction, one or the other of these theorems about tiny rings may be useful.

Construction problems

In classical construction problems (how to "construct a triangle," "mark off a segment," "draw a secant," "find a point," etc.), it is usually meant that the construction should be done with only a ruler and compass. In other words, we can draw a straight line through any two points, draw a circle of a given radius and similarly find points of intersection of lines constructed.

For the solution of such problems, *it is convenient to consider circles and straight lines as sets of points satisfying a certain condition.*

1.4. A circle is given with a point *A* outside it. Draw a straight line *l* which passes through the point *A* and is tangent to the circle.

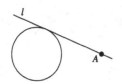

□ If *X* is the point where the straight line *l* touches the circle, then the angle *OXA* is a right angle. The set of points M for which the angle OMA is a right angle is, as we know, a circle with the diameter OA.

Thus, one can carry out the construction of the straight line l as follows. Draw a circle with the segment OA as diameter.

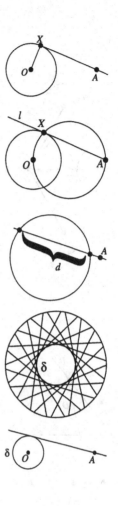

Find a point of intersection X of this circle with the given one (there are two such points symmetrical relative to the straight line OA). Finally, draw a straight line l through the points A and X. □

1.5. A point A and a circle are given. Draw a straight line through the point A so that the chord cut off by the circle along this straight line has a given length d.

□ Let us look at the set of all straight lines on which the circle marks off a chord of given length d. These straight lines are tangents to a certain circle δ whose center coincides with the center O of the given circle and whose radius is equal to $\sqrt{r^2 - d^2/4}$, where r is the radius of the given circle ⟨?⟩. The problem thus reduces to the previous one: draw through the point A a tangent to the circle δ with center O.

The problem has two solutions if the point A lies outside the circle δ: a unique solution if it lies on the circle δ, and no solution at all if it lies inside the circle δ. □

Often, it is possible to find the unknown set from the known one with the help of some simple transformations such as a *rotation*, a *symmetry*, a *parallel displacement* (or translation), or a *similarity transformation*. (This method is especially useful in construction problems.) Let us recall how we construct the image of a straight line or a circle under a translation or a similarity transformation.

For the straight line it is sufficient to plot two points A' and B', the images of two points A and B on the line, and to draw a straight line through the points A' and B'.

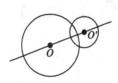

For a circle of radius r, it is sufficient to plot the point O', the image of its center O, and to draw a circle with center O' and the same radius (if the transformation is a translation) or of radius kr (if k is the ratio of magnification of a similarity transformation). We

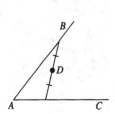

shall give some typical examples of problems where transformations are used.

1.6. A point A and a circle are given. Find the set of vertices M of the equilateral triangles ANM which have vertex N lying on the given circle. □

Let N be an arbitrary point on the given circle. If we rotate the segment AN through 60° relative to the point A, then the point N falls on the vertex M of the equilateral triangle ANM. Hence, it is obvious that if we rotate the circle as a rigid figure about the point A through an angle of 60°, then each point N of the circle will fall on the corresponding third vertex M of the equilateral triangle ANM.

Thus, all the points M lie on one of the two circles obtained from the given one by a clockwise or counterclockwise rotation about the point A through an angle of 60°.

In exactly the same way we can show that each point M on either of the two circles we obtained is the vertex of some equilateral triangle ANM. □

1.7a. An angle and a point D lying inside it are given. Construct a segment with its midpoint at the point D and its endpoints on the sides of the given angle.

□ Let us consider a set of segments which have one end lying on the side AC of the given angle (with vertex A) and their midpoint at the given point D. The other ends of these segments are obviously contained in the ray symmetric to the side AC of the angle with respect to the point D.

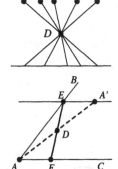

The construction reduces to the following: construct the point A' symmetric about the point A with respect to D, then draw through A' a straight line parallel to AC up to the point of intersection E with the straight line AB. Thus we obtain the required segment EF with its midpoint at D. The problem always has a unique solution.

It is interesting to note that this very construction solves the following problem.

1.7b. An angle and a point D lying inside it are given. Draw a straight line which passes through the

14

point D and which cuts off from the given angle a triangle with minimum possible area.

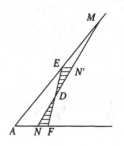

☐ We shall prove that the unknown straight line is the same straight line EF which we constructed in the previous problem, i.e., the segment between the sides of the angle that is bisected by the point D.

Let us draw through the point D a straight line MN different from EF, and prove that

$$S_{MAN} > S_{EAF}. \qquad (1)$$

We can assume that the point M on the side AB lies at a greater distance from the vertex of the angle A than E (the case when M lies closer to A than to E is analyzed similarly, interchanging the roles of the sides AB and AC). It is sufficient to verify that

$$S_{EDM} > S_{FDN}, \qquad (2)$$

as inequality (1) follows readily from this. But inequality (2) is immediate since the triangle EDM completely contains the triangle EDN' symmetric to the triangle FDN relative to the point D. ☐

Additional problems

1.8. Two points A and B are given. Find the set of feet of the perpendiculars dropped from the point A onto all possible straight lines passing through the point B.

1.9. Given a circle and a point A in a plane, find the set of midpoints of the chords cut off by the given circle on straight lines passing through the point A. (Consider all the possible cases: when the point A lies inside the circle, outside the circle, and on the circle.)

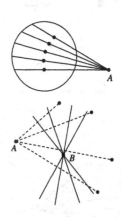

1.10. Given two points A and B, find the set of points that are symmetric about the point A with respect to some straight line passing through the point B.

1.11. Construct a circle touching two given parallel straight lines and passing through a given point which lies in between the straight lines.

1.12. Construct a circle of a given radius r touching a given straight line and a given circle.

15

1.13. A circle and two points A, B lying inside it are given. Inscribe a right-angled triangle in the circle so that the two given points lie on the sides forming the right angle. ↓

1.14. Points A and B are given. The straight line AB touches two circles, one at the point A, the other at the point B, and the circles touch each other at the point M. Find the set of such points M. ↓

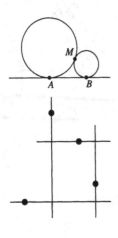

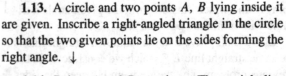

1.15. Four points are given in a plane. Find the set of centers of the rectangles formed by four straight lines passing through each of the given points. ↓

1.16. The sides OP and OQ of the rectangle $OPMQ$ lie on the sides of a given right angle. Find the set of points M in the three following cases:

(a) the length of the diagonal PQ is equal to a given value d;

(b) the sum of the lengths of the sides OP and OQ is equal to a given value d;

(c) the sum of squares of the lengths of the sides OP and OQ is equal to a given value d.

1.17. Let the rectangle $ABCD$ with diagonal length d be given; consider the sides of the rectangle as well as their extensions. Find the set of points P for which the sum of the squares of the distances from P to the four sides (or their extensions) of $ABCD$ is equal to d^2.

1.18. Let A and B be two different cities. Find the set of points M having the following property: If one travels in a straight line from M to B, then the distance from M to A is always increasing.

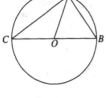

1.19. Suppose we know that in the triangle ABC the length of the median AO is:

(a) equal to half the length of the side BC;

(b) greater than half the length of the side BC;

(c) less than half the length of the side BC.

Prove that the angle A is, respectively, (a) a right angle; (b) an acute angle; (c) an obtuse angle.

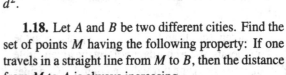

1.20. A circle and a point L are given in a plane. Find the set of midpoints of the segment LN, where N is an arbitrary point on the given circle.

16

1.21. Given a circle and a point lying outside it, draw through this point a secant such that the length of the segment of the secant outside the circle is equal to the length of the segment inside it.

1.22. Through a point of intersection of two given circles, draw a straight line on which these circles cut off chords of equal length.

1.23. Find the set of vertices C of the squares $ABCD$, where vertex A lies on a given straight line and vertex B is at a given point.

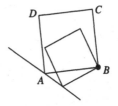

1.24. (a) Where can the fourth vertex of a square be, if two of its vertices lie on one of the sides of a given acute angle and the third vertex lies on the other side?

(b) Inscribe in a given acute triangle ABC a square, two vertices of which lie on the side AB.

1.25*. What set of points does the midpoint of the segment between two pedestrians walking uniformly along straight roads describe? (Note: There are many different answers depending on how the pedestrians are walking. Try to find all of them.) ↓

1.26*. Inside a given triangle ABC, all possible rectangles are inscribed, one side of which is on the straight line AB. Find the set of centers of all such rectangles.

1.27. A wooden right-angled triangle moves on a plane so that the vertices of its acute angles move along the sides of a given right angle. How does the vertex at the right angle of this triangle move?

1.28*. Two flat watches lie on a table. Both of them run accurately. Along what path does the midpoint M of the segment connecting the endpoints of their minute hands move? ↓

1.29*. Through the point of intersection A of two given circles, a straight line is drawn which crosses these circles once more at the points K and L, respectively. Find the set of midpoints of the segment KL. ↓

CHAPTER 2

The Alphabet

This chapter is a summary of theorems on sets or loci of points satisfying various geometric conditions. We shall gradually compile a whole list of such theorems and conditions which can be used in the solution of different types of problems.

One can draw an analogy between the geometric problem of finding a set of points and the usual algebraic problem of solving an equation (or a system of equations or an inequality). Solving an equation or an inequality means finding the set of numbers satisfying a certain condition. Just as in an algebra course, different equations (for example, trigonometric, logarithmic) are usually reduced to linear or quadratic equations, often even complicated geometric conditions turn out to be merely new properties of a straight line or circle.

The analogy between algebraic problems and problems about finding sets of points is not just a superficial one. Using analytic geometry, one type of problem can be converted into the other. Using this method, we shall see that geometric conditions, seemingly different at first glance, are covered by general theorems.

We start our geometric alphabet with the most simple assertions.

A. *The set of points equidistant from the two given points A and B is a straight line perpendicular to the segment AB and passing through its midpoint.* We shall call this straight line m the *perpendicular bisector* of the segment AB. It divides the plane into two half planes. The points in one of the half planes are closer to A than to B, and in the other closer to B than to A. The points A and B are symmetric relative to m.

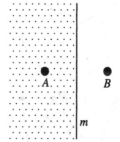

19

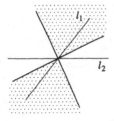

B. *The set of points equidistant from two given intersecting straight lines l_1 and l_2 is two mutually perpendicular straight lines which bisect the angles formed by the straight lines l_1 and l_2.*

These straight lines are the axes of symmetry of the figure formed by the straight lines l_1 and l_2. This set— *the cross bisector*—divides the plane into four regions. In the figure two right angles—the set of points closer to the straight line l_1 than to the line l_2—are shown.

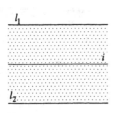

C. *Given a straight line l and a positive number h, the set of points at a distance h from l is a pair of straight lines l_1, l_2, parallel to l and lying on opposite sides of l.*

The belt between the straight lines l_1 and l_2 is the set of points which are at a distance less than h from the straight line l.

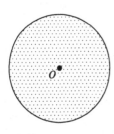

D. *Given a point O and a positive number r, the set of points at a distance r from O is a circle of radius r with center O.*

(This is the definition of a circle.)

The circle divides the plane into two parts: the region inside and the region outside. For points inside the circle, the distance from the center is less than r and for points outside the circle it is greater than r.

We give a few simple reformulations of the conditions **A**, **B**, and **C** in the form of the following four problems.

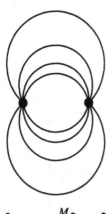

2.1. Find the set of centers of the circles passing through two given points.

2.2. Find the set of centers of the circles touching two given intersecting straight lines.

2.3. Find the set of centers of circles of radius r touching a given straight line.

2.4. Given two points A and B, find the set of points M for which the area S_{AMB} of the triangle AMB is equal to a given number $c > 0$.

We illustrate Proposition **B** with a less trivial example—we prove the theorem on the bisector of a triangle.

20

2.5. Let the cross bisector of the straight lines AC and BC intersect the straight line AB at the points E and F. Prove that

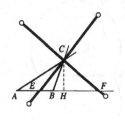

$$\frac{|AE|}{|EB|} = \frac{|AF|}{|FB|} = \frac{|AC|}{|CB|}.$$

□ Let M be one of the points E or F. Note that

$$\frac{|AM|}{|MB|} = \frac{S_{ACM}}{S_{MCB}}.$$

(The triangles ACM and MCB have the same height CH.)

The ratio of the areas can also be expressed in a different way; since the point M belongs to the cross bisector, it is equidistant from the straight lines AC and BC, hence,

$$\frac{S_{ACM}}{S_{MCB}} = \frac{|AC|}{|CB|}. \quad □$$

A circle and a pair of arcs

The next step of our "alphabet" is one more variant of the theorem on the inscribed angle and the ring on a circle which we discussed in Chapter 1.

E°. *Two intersecting straight lines l_A and l_B rotate in the same plane about two of their points A and B with the same angular velocity ω* (here, the value of the angle between them obviously remains constant). *The trajectory of the point of intersection of these straight lines is a circle.*

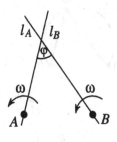

□ Construct a circle δ passing through three points: A, B and a particular position M_0 of the point of intersection of the straight lines l_A and l_B. According to the theorem about the tiny ring on a circle given in Chapter 1, the point of intersection of the straight line l_A and the circle δ moves uniformly along the circle δ with angular velocity 2ω. The point of intersection of l_B with the circle δ moves in exactly the same way. As they are coincident at a particular instant (at the position M_0), they also coincide at any other instant of time. □

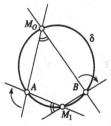

We shall give an alternative variant of Theorem **E** without using the language of motion.

21

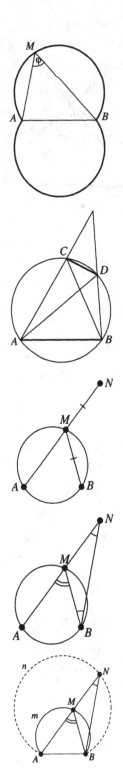

E. *The set of points at which the given segment AB subtends an angle of given value φ (i.e., the set of points M for which $\widehat{AMB} = \varphi$) is a pair of arcs with their endpoints at A and B which are symmetric about the straight line AB.*

The region bounded by these two arcs is a set of points M for which $\widehat{AMB} > \varphi$.

Note that if $\varphi = 90°$, then the set **E** will be a circle with diameter AB. We have already mentioned this following Problem **0.1**.

2.6. Chord AB of a given circle is fixed, and another chord, CD, is moved along the circle without altering its length. Along what path does the point of intersection of the lines (a) AD and BC, (b) AC and BD move?

2.7. Given two points A and B, find the set of vertices M and N of parallelograms $AMBN$ with the given angle $\widehat{MAN} = \varphi$.

2.8a. A circle and two points A and B on it are given. Let M be an arbitrary point on this particular circle. A segment MN equal to the segment BM in length is marked off from the point M on the segment AM produced. Find the set of points N.

□ Let N be some point plotted as stated in the problem. Then $|BM| = |NM|$ and $\widehat{NBM} = \widehat{MNB}$. But since $\widehat{AMB} = \widehat{MBN} + \widehat{MNB}$, then $\widehat{ANB} = \widehat{AMB}/2$. The value of the angle AMB for all points M lying on one of the arcs $\overset{\frown}{AmB}$ is constant (see **E**): $\widehat{AMB} = \varphi$. Hence $\widehat{ANB} = \varphi/2$, i.e., all these points lie on the arc $\overset{\frown}{AnB}$ containing the angle $\varphi/2$. (The center of the arc lies at the midpoint of the arc $\overset{\frown}{AmB}$ of the given circle (?).)

Do all the points of the arc $\overset{\frown}{AnB}$ satisfy our requirements? No, not all of them.

Note that when the point M runs along the arc $\overset{\frown}{AnB}$ from the point B to the point A, the chord AM rotates about the point A from the straight line AB up to the tangent to the given circle at the point A. Hence, only

22

a part of the arc $\overset{\frown}{AnB}$ and in particular the arc $\overset{\frown}{EnB}$ (where E is the point of intersection of the arc $\overset{\frown}{AnB}$ with the tangent at the point A) belongs to the set we are looking for.

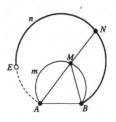

Note that we can take the point B as belonging to our set (when M coincides with B the "length" of the segment MB is equal to zero). Strictly speaking, the point E does not belong to our set; when the point M coincides with the point A, the direction of the straight line AM has no meaning.

The points lying on the other side of the line AB are treated in a similar way.

Thus, the unknown set of points consists of two arcs

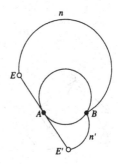

$\overset{\frown}{EnB}$ and $\overset{\frown}{E'n'B}$. $\square$

We may solve Problem **2.8a** in a different way, if we notice that the points N and B are symmetric about the straight line CM, where C is the midpoint of the

arc $\overset{\frown}{AmB}$. From this it follows that the set of points N reduces to the set of points found in Problem **1.10** for the points A and C.

We present a problem similar to **2.8a** for the reader to examine in the same way.

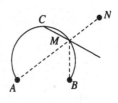

2.8b. Solve Problem **2.8a**, but now assume that the segment MN is marked off in the opposite direction on the ray MA.

Squares of distances

Consider two points A and B in a plane and an arbitrary number c.

F. *The set of points M for which*

$$|AM|^2 - |BM|^2 = c$$

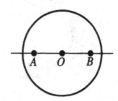

is a straight line perpendicular to the segment AB (in particular, when $c = 0$, we get the perpendicular bisector; see Proposition **A**).

G. Suppose $|AB| = 2a$. The *set of points M for which*

$$|AM|^2 + |BM|^2 = c$$

is:

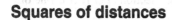

(a) *a circle with its center* at the midpoint O of the segment AB and of radius $r = \sqrt{(c - 2a^2)/2}$, when $c > 2a^2$;

(b) a *point* O, when $c = 2a^2$;

(c) the *empty set*, when $c < 2a^2$.

It is not difficult to prove Propositions **F** and **G** analytically, using (x, y)-coordinates, or with the Pythagorean Theorem $\langle ? \rangle$.

We shall not present a separate proof for each statement, but deduce them both as corollaries of a more general theorem. But first we illustrate them with a few examples.

2.9. Find the set of points for which the tangents drawn to two given circles are equal in length.

□ Let O_1 and O_2 be the centers of the given circles, r_1 and r_2 their radii ($r_2 \geq r_1$), and let MT_1 and MT_2 be the tangents to them drawn from the point M. Using the Pythagorean Theorem, the condition $|MT_1|^2 = |MT_2|^2$ may be written as

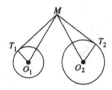

$$|MO_1|^2 - |O_1T_1|^2 = |MO_2|^2 - |O_2T_2|^2,$$

or

$$|MO_2|^2 - |MO_1|^2 = r_2^2 - r_1^2.$$

According to Proposition **F**, the set of points M belongs to the straight line perpendicular to O_1O_2.

If the circles intersect, this straight line will pass through their points of intersection. For if A is one of these points, then

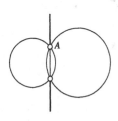

$$|O_2A|^2 - |O_1A|^2 = r_2^2 - r_1^2$$

and, consequently, the point A lies on this straight line. The required set of points in this case is shown in the figure; it is the union of two rays.

If the circles are concentric (and $r_2 > r_1$), the required set is empty. If the circles coincide, the set consists of all the points outside the circle. If the circles are nonintersecting and nonconcentric, the answer will be a straight line. □

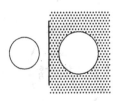

The straight line discussed in Problem **2.9** is called the *radical axis of the two circles*. Suppose two nonintersecting circles are given. Then their radical

24

axis divides the complement of the two circles into two regions: the set of points M for which $|MT_1| > |MT_2|$ and the set of points M for which $|MT_1| < |MT_2|$.

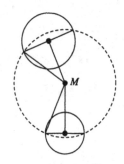

2.10. Find the set of centers of the circles which intersect each of two given circles at diametrically opposite points.

2.11. (a) The sum of the squares of the lengths of the diagonals of a parallelogram is equal to the sum of squares of the lengths of its sides. Prove this.

(b) If the diagonals of a convex quadrilateral $AMBN$ are mutually perpendicular, then $|AM|^2 + |BN|^2 = |AN|^2 + |BM|^2$. Prove this. ↓

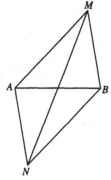

□ (a) Let the vertices A and B of the parallelogram $AMBN$ be at a distance a from its center O, let the vertices M and N at a distance r from O, and let $c = 2(a^2 + r^2)$. Consider the figure opposite, where O is marked as the center of the parallelogram. In this figure, $|OM| = \sqrt{(c - 2a^2)/2}$, so according to Proposition G, the sum of the squares of the distances from the point M to the points A and B is equal to c. In the same way $|AN|^2 + |BN|^2 = c$; hence

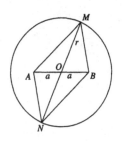

$$|AM|^2 + |BM|^2 + |AN|^2 + |BN|^2$$
$$= 2c = 4(a^2 + r^2) = |MN|^2 + |AB|^2. \quad \square$$

We now present the general theorem which contains Propositions **F, G, A, D** of our alphabet.

Theorem on the Squares of the Distances. The set of points M for which the condition

$$\lambda_1|MA_1|^2 + \lambda_2|MA_2|^2 + \cdots + \lambda_n|MA_n|^2 = \mu \quad (1)$$

is satisfied, where A_1, A_2, ..., A_n are given points, λ_1, λ_2, ..., λ_n, μ are given numbers, is one of the following simple geometric figures:

1°. *If $\lambda_1 + \lambda_2 + \cdots + \lambda_n \neq 0$, it will be a circle, a point, or the empty set.*

2°. *If $\lambda_1 + \lambda_2 + \cdots + \lambda_n = 0$, it will be a straight line, the entire plane, or the empty set.*

We shall give a proof of the theorem using analytic methods.

□ The square of the distances between the points $M(x, y)$ and $A_k(x_k, y_k)$ is calculated according to the

25

formula

$$|MA|^2 = (x - x_k)^2 + (y - y_k)^2$$
$$= x^2 + y^2 - 2x_k x - 2y_k y + x_k^2 + y_k^2.$$

Consider the expression

$$\lambda_1 |MA_1|^2 + \lambda_2 |MA_2|^2 + \cdots + \lambda_n |MA_n|^2.$$

In order to write it in coordinates, it is necessary to sum together several expressions of the form

$$\lambda(x^2 + y^2 - 2px - 2qy + p^2 + q^2).$$

As a result, condition (1) may be written in the form of the equation

$$dx^2 + dy^2 + ax + by + c = 0, \qquad (2)$$

where $d = \lambda_1 + \lambda_2 + \cdots + \lambda_n$.

We shall now prove that *equation (2) gives one of the figures enumerated above.*

1°. If $d \neq 0$, we can transform (2) in the following manner:

$$x^2 + y^2 + \frac{a}{d}x + \frac{b}{d}y + \frac{c}{d} = 0$$

or

$$\left(x + \frac{a}{2d}\right)^2 + \left(y + \frac{b}{2d}\right)^2 = \frac{b^2 + a^2 - 4dc}{4d^2}. \qquad (2')$$

We can see that this gives us:

a circle with center at the point $C(-a/2d, -b/2d)$, if the right-hand side of (2') is positive;

a single point $C(-a/2d, -b/2d)$, if the right-hand side equals zero;

the empty set, if the right-hand side is negative.

2°. If $d = 0$, equation (2) takes the form

$$ax + by + c = 0.$$

This will be:

a straight line, if $a^2 + b^2 \neq 0$,

the entire plane, if $a = b = c = 0$,

the empty set, if $a = b = 0$, $c \neq 0$. □

As a rule, in a particular example, it is easy to determine which of these cases is involved. Let us return again to Propositions **F** and **G** of our "alphabet" which have not been proved yet.

Proof of F. The condition $|MA|^2 - |MB|^2 = c$ is a particular case of (1), where $n = 2$, $\lambda_1 = 1$, $\lambda_2 = -1$, from which $d = 0$, and hence it determines either a straight line or a plane, or the empty set.

Since the equation $(x + a)^2 - (x - a)^2 = c$ always has a single solution, $x = c/4a$, one point of the set is on the straight line AB. Therefore, the required set is a straight line. It is clear from symmetry considerations that this straight line is perpendicular to the straight line AB. $\square$

Proof of G. The condition $|MA|^2 + |MB|^2 = c$ is a particular case of (1). Here $\lambda_1 = 1$, $\lambda_2 = 1$, $d \neq 0$, and, therefore, the unknown set would be either the empty set, a point, or a circle. Since the points A and B appear in the condition symmetrically, the center of the circle lies at the midpoint of the segment AB.

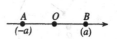

In order to find when the unknown set is a circle and to determine its radius, we find the points on the straight line AB which satisfy the condition $|AM|^2 + |BM|^2 = c$. To do this, note that the equation $(x-a)^2 + (x+a)^2 = c$ has a solution when $c \geq 2a^2$, and

$$x \mid = r = \sqrt{(c - 2a^2)/2}. \ \square$$

2.12. Let a rectangle $ABCD$ be given in the plane. Find the set of points M in the same plane for which $|MA|^2 + |MC|^2 = |MB|^2 + |MD|^2$.

$\square$ *Answer:* The entire plane. Let us prove this. Let $ABCD$ be as shown in the figure. Then we seek the set of points M for which $|MA|^2 + |MC|^2 - |MB|^2 - |MD|^2 = 0$.

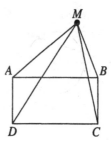

In condition (1) (p. 25) put $n = 4$, $\lambda_1 = \lambda_2 = 1$, $\lambda_3 = \lambda_4 = -1$ and $\lambda_1 + \lambda_2 + \lambda_3 + \lambda_4 = 0$. According to the theorem, the required set is either a straight line, or the empty set, or the entire plane.

We note that the vertices A, B, C, D of the rectangle itself satisfy the condition. For example, the following equality $|AA|^2 + |AC|^2 - |AB|^2 - |AD|^2 = 0$ (the Pythagorean Theorem) is valid for the point A. Thus,

27

the required set is neither the empty set nor a straight line. Hence, it follows that the required set is the entire plane. □

From the result of Problem **2.12**, it follows that if $ABCD$ is a rectangle, then for any point M of the plane the following equality holds:

$$|MA|^2 + |MC|^2 = |MB|^2 + |MD|^2.$$

Solve the following problem using this fact.

2.13. A circle and a point A inside it are given. Find the set of the fourth vertices C of the rectangles $ABCD$, whose vertices B and D belong to the given circle.

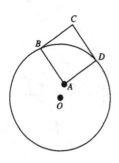

In the next problem, we introduce the notation $\rho(M, m)$ to denote the distance between the point m and the line M.

2.14. Prove that $|MA|^2 - |MB|^2 = 2|AB|\rho(M, m)$, where m is the perpendicular bisector of the segment AB, and $|MA| > |MB|$.

We add to our alphabet one more proposition which is frequently used in geometry and is also a corollary from the theorem on the squares of the distances.

H. *The set of points M for which*

$$|MA|/|MB| = k, \quad k > 0, \quad k \neq 1,$$

is a circle whose diameter belongs to the straight line AB.

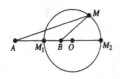

Given any point M in the above set, the ratio of the distance from M to A to the distance from M to B is a constant. This set of points M is called the *circle of Apollonius*.

□ Let us rewrite condition **H** in the form

$$|MA|^2 - k^2|MB|^2 = 0.$$

This condition is a particular case of the condition (1) (p. 25) where $n = 2$, $\lambda_1 = 1$, $\lambda_2 = -k^2$ and hence if $1 - k^2 \neq 0$, the required set will be a circle, a point, or the empty set. Since the equation

$$(x + a)^2 = k^2(x - a)^2$$

always has two solutions when $k^2 \neq 1$, there exist two points M_1 and M_2 that lie in the intersection of this set and the straight line AB. Hence the unknown set is a circle. As the condition is symmetric relative to the straight line AB, the diameter of this circle is the segment $M_1 M_2$. $\square$

Incidentally, note that if M is a point of the circle of Apollonius, then the cross bisector of the straight lines AM and MB intersects the line AB at the points M_1 and M_2. (This follows from the theorem on the cross bisector in **2.5**, since $|AM_1|/|BM_1| = |AM_2|/|BM_2| = |AM|/|BM|$.)

This argument is used in the next problem.

2.15. Two billiard balls A and B are placed on the diameter of a circular billiard table. Ball B is hit in such a way that after one rebound from the side of the table, it strikes ball A. Find the trajectory of ball B if the stroke is not directed along the diameter.

2.16. The points A, B, C, D are on a given straight line. Construct a point M in the plane from which the segments AB, BC and CD are seen at one and the same angle (i.e., subtend the same angle at M).

Distances from straight lines

So far in this chapter, we have mainly used various properties that define a circle. In the next two propositions of our alphabet, we will see only pairs of straight lines.

We shall consider two intersecting straight lines l_1 and l_2 in a plane and a positive number c.

I. *The set of points M, the ratio of whose distances from the straight lines l_1 and l_2 is equal to a constant c (that is, $\rho(M, l_1)/\rho(M, l_2) = c$), is a pair of straight lines passing through the point of intersection of the straight lines l_1 and l_2.*

J. *The set of points M, the sum of whose distances from the straight lines l_1 l_2 is equal to a constant c (that is, $\rho(M, l_1) + \rho(M, l_2) = c$), is the boundary of a rectangle with diagonals lying on the lines l_1 and l_2.*

Before proving these theorems, let us illustrate their application in the following two examples.

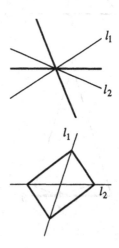

2.17. Given a triangle ABC, find the set of all points M for which $S_{AMC} = S_{BMC}$.

□ Let h_a and h_b be, respectively, the distances of the point M from the straight lines AC and BC. Then,

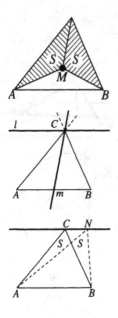

$$S_{AMC} = \frac{|AC| \cdot h_b}{2}, \quad S_{BMC} = \frac{|BC| \cdot h_a}{2},$$

consequently $h_a/h_b = |AC|/|BC|$.

Hence the required set of points M is the set given in Proposition **I** for the lines AC and BC and $c = |AC|/|BC|$. Thus it represents a pair of straight lines passing through the point C. We shall show that one of the straight lines m contains the median of the triangle, and the other, l, is parallel to the straight line AB. For this, it is sufficient to take a single point on each of the straight lines and verify that the stated condition is fulfilled for them.

Let us denote by h the altitude of the triangle drawn from the vertex C. Let N be a point on the straight line l. Then

$$S_{ACN} = \frac{|CN| \cdot h}{2} \text{ and } S_{BCN} = \frac{|CN| \cdot h}{2}.$$

Hence $S_{ACN} = S_{BCN}$ and the straight line l belongs to the required set.

Let K be the midpoint of the side AB, i.e., $|AK| = |KB|$. Then $S_{AKC} = |AK| \cdot h/2 = |BK| \cdot h/2 = S_{BKC}$, and consequently, the whole line m belongs to the unknown set. □

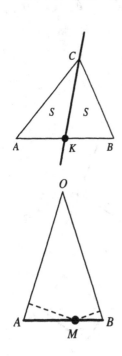

In analogy with the cross bisector, one may call the pair of straight lines m and l "the cross median" of the vertex C of the triangle.

Proposition **J** can, in essence, be reduced to the following problem.

2.18. Given an isosceles triangle AOB, prove that the sum of the distances from the point M on its base AB to the straight lines AO and BO is equal to the length of the altitude dropped from A onto the side BO.

We shall not give geometrical proofs of Propositions **I** and **J**, although they are not at all difficult. But we shall give proofs using the language of motion. (As

30

was done above in Proposition $\mathbf{E}°$ "A circle and a pair of arcs.") Let us first formulate a lemma which generalizes the theorem on a tiny ring on a straight line (p. 11).

Lemma. A tiny ring M is placed on two straight lines l_1 and l_2 at their point of intersection. If each straight line describes a uniform translatory motion, then the ring M moves uniformly along some straight line.

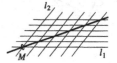

☐ This straight line can be constructed by marking two different positions M_1 and M_2 of the ring. The points of intersection of the moving straight lines with the stationary line $M_1 M_2$ move uniformly. Since these points coincide with each other at two different instances in time (when the ring M passes through M_1 and M_2), they must always coincide. ☐

Proof of I. The set of points lying at a distance t from l_2 and a distance ct from l_1 for some positive number t consists of the four vertices of a parallelogram whose center is at the point O of intersection of l_1 and l_2. For the set of points lying at a distance t from l_2 is a pair of parallel lines (see $\mathbf{C}$) and the set of points lying at a distance ct from l_1 is also a pair of parallel lines; their points of intersection are the four vertices of the parallelogram. These four points satisfy the condition stated in $\mathbf{I}$ since

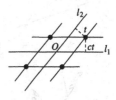

$$ct/t = c.$$

By varying the number t from zero to infinity, we get all the points of the required set.

By regarding t as "time," we see that the four straight lines constructed above move uniformly (remaining parallel to l_1 and l_2). By the lemma, their points of intersection, the rings, move along a straight line passing through the point O. ☐

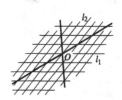

Proof of J. Draw two straight lines at a distance t from l_1 and two at a distance $c - t$ from l_2 ($0 \leq t \leq c$). The four points of intersection of these straight lines belong to the required set. When the "time" t varies from zero to c, the straight lines move uniformly and each of the four points of intersection, by the lemma, moves through a segment. The endpoints of these segments,

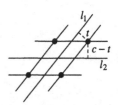

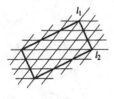

which correspond to $t = 0$ and $t = c$, lie on the straight lines l_1 and l_2 and are the vertices of a rectangle. $\square$

We shall now state a general theorem which includes Propositions **B, C, I, J** of the alphabet. Recall that $\rho(M, m)$ represents the distance between the point m and the line M. Consider *the set of points M for which*

$$\lambda_1\rho(M, l_1) + \lambda_2\rho(M, l_2)$$
$$+ \cdots + \lambda_n\rho(M, l_n) = \mu. \tag{3}$$

Here $l_1, l_2, \ldots, l_n$ are given straight lines, and $\lambda_1, \lambda_2, \ldots, \lambda_n, \mu$ are given numbers.

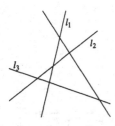

It is difficult to give an immediate description of this set in the entire plane. However, as we shall now see, in each of the pieces into which the straight lines l_1, $l_2, \ldots, l_n$ divide the plane, set (3) is, as a rule, simply a part of some straight line. Let us denote one of these pieces by Q.

Theorem on the Distances from the Straight Lines.
The set of points which satisfy condition (3), belonging to Q, is either (1) the intersection of Q with a straight line, i.e., a ray, a segment, or even a whole straight line, or (2) the whole of Q, or (3) the empty set.

By finding the set on each of the pieces, we shall find the entire required set (as in **1.3**). We shall give an analytic proof of the theorem.

$\square$ Suppose we want to find the set of points on one of the pieces Q of the plane into which the lines l_1, $l_2, \ldots, l_n$ divide the plane. The piece Q of the plane can be imagined as the intersection of n half planes with boundary lines $l_1, l_2, \ldots, l_n$.

The equation $a_k x + b_k y + c_k = 0$ of the straight line l_k can be selected such that inside the required half plane, $a_k x + b_k y + c_k \geq 0$ and $a_k^2 + b_k^2 = 1$ ⟨?⟩; then for the point $M(x, y)$ in this half plane, $\rho(M, l_k) = a_k x + b_k y + c_k$.

In order to write the quantity $\lambda_1\rho(M, l_1) + \lambda_2\rho(M, l_2) + \cdots + \lambda_n\rho(M, l_n)$ in coordinates, we have to add several linear expressions of the form $\lambda_k a_k x + \lambda_k b_k y + \lambda_k c_k$. As a result, condition (3) is expressed by a linear equation

$$ax + by + c = 0.$$

If $a^2 + b^2 \neq 0$, this equation represents a straight line. If $a = b = 0$, it represents either the entire plane or the empty set. □

One can obtain an alternative proof of this theorem by using Problem **2.14** to reduce the theorem to the previous theorem on the squares of the distances (given on p. 25). ⟨?⟩.

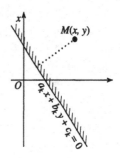

2.19. (a) A right triangle ABC is given. Find the set of points for which the sum of the distances from the straight lines AB, BC and CA is equal to a given number $\mu > 0$. ↓

(b) Given a rectangle $ABCD$. Find the set of points for which the sum of the distances from the straight lines AB, BC, CD, DA is equal to a given number μ.

2.20.* (a) Three straight lines l_0, l_1, l_2 intersect at a single point. The value of the angle between any two of them is equal to $60°$. Find the set of points M for which

$$\rho(M, l_0) = \rho(M, l_1) + \rho(M, l_2).$$

(b) An equilateral triangle ABC is given. Find the set of points M whose distance from one of the straight lines AB, BC, CA is half the sum of its distances from the remaining two lines. ↓

The entire "alphabet"

The set of points satisfying a certain condition is denoted as follows: Inside the braces a letter is first written to denote an arbitrary point of the set (in our case, it is, as a rule, the letter M, but it can be any letter); then there is a colon which is followed by the condition which specifies the required set of points.

Let us now summarize the sets of our "alphabet":

A. $\{M : |MA| = |MB|\}$.
B. $\{M : \rho(M, l_1) = \rho(M, l_2)\}$.
C. $\{M : \rho(M, l) = h\}$.
D. $\{M : |MO| = r\}$.
E. $\{M : \widehat{AMB} = \varphi\}$.
F. $\{M : |AM|^2 - |MB|^2 = c\}$.

33

G. $\{M : |AM|^2 + |MB|^2 = c\}$.

H. $\{M : |AM|/|MB| = k\}$.

I. $\{M : \rho(M, l_1)/\rho(M, l_2) = k\}$.

J. $\{M : \rho(M, l_1) + \rho(M, l_2) = c\}$.

Recall that we have separated the propositions of our "alphabet" with the exception of **E** into two groups:

A, D, F, G, H and **B, C, I, J.**

The sets in the first group are particular cases of the set

$$\{M : \lambda_1|MA|^2 + \lambda_2|MA_2|^2 + \cdots + \lambda_n|MA_n|^2 = \mu\},$$

and the sets in the second group are particular cases of the set

$$\{M : \lambda_1\rho(M, l_1) + \lambda_2\rho(M, l_2) + \cdots + \lambda_n\rho(M, l_n) = \mu\}.$$

In Chapter 6 we shall add four more "letters" to our "alphabet":

K. $\{M : |MA| + |MB| = c\}$.

L. $\{M : ||MA| - |MB|| = c\}$.

M. $\{M : |MA| = \rho(M, l)\}$.

N. $\{M : |MA|/\rho(M, l) = c\}$.

These sets are ellipses, hyperbolas, and parabolas. These curves also fall naturally into a single group known as the *quadratic curves*.

34

CHAPTER 3

Logical Combinations

In this chapter we have collected various problems which, as a rule, involve combinations of several geometric conditions. In solving these problems, we will learn to classify points and to consider logical relations between conditions as operations on sets.

Through a single point

In the first few problems, we touch on the traditional subject matter of geometry. With the help of simple manipulations using the sets of our "alphabet," we will prove some theorems on various special points associated to a triangle. The central logic underlying much of our reasoning depends on the fact that equality is a transitive property: if $a = b$ and $b = c$, then $a = c$.

3.1. In a triangle ABC the midperpendiculars (perpendicular bisectors of the sides; see p. 19) intersect at a single point. This point is the center of the circumscribed circle of the triangle (the circumscribed circle is also known as the *circumcircle*).

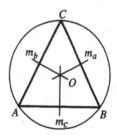

☐ The midperpendiculars m_c and m_a of the sides AB and BC must intersect at some point O. Since the point O belongs to the midperpendicular m_c, then according to **A** (Chapter 2), the equality $|OA| = |OB|$ holds true. In exactly the same way, the fact that O belongs to the midperpendicular m_a implies that $|OB| = |OC|$. Hence $|OA| = |OC|$ and consequently the point O belongs to the midperpendicular m_b of the side AC.

We have thus proved that all three midperpendiculars intersect at the point O. ☐

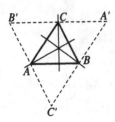

3.2. Recall that an *altitude* of a triangle is the line that passes through a given vertex and is perpendicular to the opposite side. The three altitudes of a triangle ABC intersect at a single point, called the *orthocenter* of the triangle.

□ Through each of the vertices of the triangle, draw a straight line parallel to the side opposite the vertex. These straight lines form a new triangle $A'B'C'$. The points A, B, and C are the midpoints of the sides of the new triangle $A'B'C'$; the altitudes of the triangle ABC belong to the perpendicular bisectors of the sides $A'B'$, $B'C'$, and $C'A'$. Hence by **3.1**, they are concurrent. □

We shall give a second proof of **3.2**, similar to that of **3.1**.

□ Let us consider each of the altitudes as a set of points satisfying a certain condition. For this we shall use Proposition **F** of the "alphabet."

We know that the set

$$\{M: |MA|^2 - |MB|^2 = d\}$$

is a straight line perpendicular to AB. Choose d such that this straight line contains the vertex C. To do this, we must take $d = |CA|^2 - |CB|^2$. Thus, the straight line

$$h_c = \{M: |MA|^2 - |MB|^2 = |CA|^2 - |CB|^2\}$$

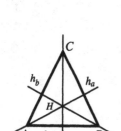

contains the altitude of the triangle dropped from the vertex C.

One can consider the straight lines containing two other altitudes of the triangle in a similar way.

$$h_a = \{M: |MB|^2 - |MC|^2 = |AB|^2 - |AC|^2\},$$
$$h_b = \{M: |MC|^2 - |MA|^2 = |BC|^2 - |BA|^2\}.$$

Suppose the first two straight lines h_c and h_a intersect at the point H. Then when M coincides with this point, both of the following equations hold:

$$|HA|^2 - |HB|^2 = |CA|^2 - |CB|^2,$$
$$|HB|^2 - |HC|^2 = |AB|^2 - |CA|^2.$$

Adding these two equalities, we obtain

$$|HA|^2 - |HC|^2 = |AB|^2 - |CB|^2.$$

36

Hence, the point H also belongs to the third straight line h_b. $\square$

3.3. Three bisectors of the angles of a triangle ABC intersect at a single point at the center of the inscribed circle of the triangle. (The inscribed circle of the triangle is also known as the *incircle* of the triangle.)

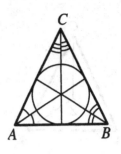

$\square$ Let a, b and c be the straight lines to which the sides of the triangle belong. The bisectors l_a and l_b of the angles A and B must intersect at some point O (inside the triangle). For this point O, the following equalities hold (note that these follow from Theorem B of our alphabet):

$$\rho(O, b) = \rho(O, c) \quad \text{and}$$
$$\rho(O, a) = \rho(O, c).$$

Hence, $\rho(O, b) = \rho(O, a)$ and point O belongs to the bisector l_c of angle C of the triangle. $\square$

Note. The set of points M in the plane for which $\rho(M, c) = \rho(M, b)$ and $\rho(M, a) = \rho(M, c)$ consists of four points: O, O_1, O_2 and O_3, the points of intersection of the two cross bisectors. Reasoning similarly as in the solution of **3.3**, we find that the third cross bisector (the cross bisector of the straight lines a and b) also passes through these points.

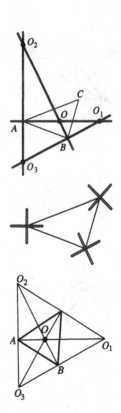

From here it follows that the six bisectors of the internal and external angles of the triangle intersect in threes at four points. One of these points is the center of the inscribed circle and the other three are the centers of the so-called *escribed circles*.

Note that, if in an arbitrary acute-angled triangle $O_1 O_2 O_3$ the points A, B, C are the feet of its altitudes, then O_1, O_2 and O_3 are the centers of the escribed circles (sometimes called *excircles*) of the triangle ABC. The altitudes of the triangle $O_1 O_2 O_3$ are therefore the bisectors of the angles of the triangle ABC.

3.4. The medians of a triangle intersect at a single point, called the centroid of the triangle (or the *center of gravity* of the triangle).

This theorem can be proved by different methods.

The first proof, which we give here, explains the term "the center of gravity" of the triangle.

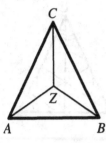

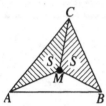

□ Let us place three weights W_A, W_B, W_C of the same mass, say 1 g, at the vertices of the triangle ABC, and find the position of their center of gravity. The center of gravity of the two weights W_A and W_B lies at the midpoint of the segment AB; hence, the center of gravity Z lies on the corresponding median. We can show in the same way that Z belongs to the other two medians. Hence, all three medians intersect at the point Z. □

We shall also give a proof along the same lines as the three previous proofs.

□ Suppose we are given a triangle ABC. The points of the medians of the triangle down from the vertices A, B, C satisfy the following conditions (respectively) (see **2.17**):

$$S_{AMB} = S_{CMA}, \qquad (1)$$
$$S_{AMB} = S_{BMC}, \qquad (2)$$
$$S_{BMC} = S_{CMA}.$$

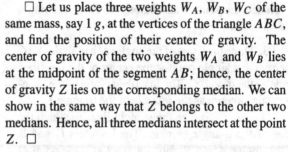

It is clear that the third condition follows from the first two, and so the medians intersect at a single point Z. □

Note. The set of points that satisfy the conditions of equation (1) is, according to **2.17**, a pair of straight lines which we call the "cross median." Thus, three such sets intersect at four points: Z, A', B', C'. Note that the triangle $A'B'C'$ is just the triangle considered in the first proof of the theorem on the altitudes in **3.2**.

3.5. (a) Prove that for any three circles, the three radical axes of the pairs of circles either pass through a single point or are parallel (see **2.9**).

(b). Prove that if three circles intersect in pairs, then the three common chords of each pair of circles (or their extensions) pass through a single point or are parallel. ↓

3.6. (Torricelli's point). Prove that in an acute-angled triangle ABC, there exists a point T (Torricelli's point) at which all the sides subtend the same angle (i.e., such that $\widehat{ATB} = \widehat{BTC} = \widehat{CTA}$).

3.7. Consider all the possible triangles with a given base AB with the vertex angle equal to φ. Find the set of:

    (a) points of intersection of the medians,

    (b) points of intersection of the angle bisectors, ↓

    (c) points of intersection of the altitudes. ↓

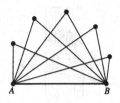

3.8. (a) Three straight lines a, b, c (intersecting in pairs) pass through three given points A, B, C, respectively. The lines rotate with angular velocity ω. Prove that at some point in time these straight lines pass through a single point. ↓

    (b) Prove that three circles symmetric to the circumcircle of the triangle ABC relative to the straight lines AB, BC and CA pass through a single point, the orthocenter of the triangle ABC. ↓

3.9. (Ceva's Theorem). Points C_1, A_1, B_1 are selected on the sides AB, BC, CA of the triangle. Prove that the segments AA_1, BB_1 and CC_1 are concurrent (intersecting at a single point) if and only if the condition

$$\frac{|AC_1|}{|C_1B|} \cdot \frac{|BA_1|}{|A_1C|} \cdot \frac{|CB_1|}{|B_1A|} = 1$$

is satisfied. ↓

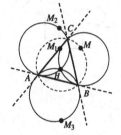

3.10. Suppose we are given a triangle ABC with sides AB, BC, and CA. Let the points C_1, A_1, B_1 lie on the sides AB, BC, CA, respectively, and suppose that at each of those points, we erect a perpendicular to that side.

Prove that these three perpendiculars are concurrent if and only if the condition

$$|AC_1|^2 + |BA_1|^2 + |CB_1|^2$$
$$= |AB_1|^2 + |BC_1|^2 + |CA_1|^2$$

is satisfied. ↓

Intersections and unions

We now single out the basic operations that will we be using constantly.

Suppose two or more sets of points are given. The set of all points belonging simultaneously to all the

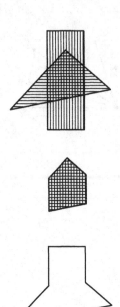

given sets is called the *intersection* of the sets. The set of all points belonging to at least one of the given sets is called the *union* of these sets.

When a problem requires us to find those points that *simultaneously* satisfy *several conditions*, we find the set of points satisfying each of the conditions separately and then take the intersection of these sets. We meet a similar situation in algebraic problems as well. The set of solutions of the system of equations

$$\begin{cases} f_1(x) = 0 \\ f_2(x) = 0 \end{cases}$$

is in fact the intersection of the solution sets of the individual equations making up this system.

When a problem requires us to find those points that satisfy *at least one* of several conditions, we find the set of points satisfying separately each of the conditions and then take the union of these sets. This is what we do, for example, when solving the equation $f(x) = 0$ when the left-hand side may be factorized as

$$f(x) = f_1(x) f_2(x).$$

We find the solution set for each of the equations $f_1(x) = 0$, $f_2(x) = 0$ and then take their union.

There is another concept which gives rise to an algebraic association: namely, the partition (or subdivision) of a domain. In order to solve the inequality $f(x) > 0$ or $f(x) < 0$, it is usually sufficient to solve the corresponding equation $f(x) = 0$. The points obtained divide the domain of definition of the function f (an interval or the whole line) into pieces, in each of which the function does not change sign. In exactly the same way, the sets of points of a plane for which various inequalities hold are usually domains that are themselves bounded by the lines on which the corresponding equalities are satisfied. We have already seen many simple examples of this type in Chapter 2.

We shall encounter more complicated partitions and combinations of sets in the next problem.

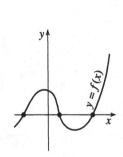

3.11. Let two points A and B be given in a plane. Find the set of points for which the triangle AMB is:

(a) a right-angled triangle,

(b) an acute-angled triangle,

(c) an obtuse-angled triangle.

□ (a) The triangle AMB is a right-angled triangle if one of the following three conditions is met: (1) $\widehat{AMB} = 90°$, (2) $\widehat{BAM} = 90°$, (3) $\widehat{ABM} = 90°$.

The unknown set is, therefore, the union of the following three sets: (1) a circle with $|AB|$ as diameter, (2) a straight line l_A passing through the point A and perpendicular to the segment AB, (3) a straight line l_B passing through the point B and perpendicular to the segment AB.

We must exclude from this union the points A and B on the line AB, as they give rise to a "degenerate" triangle AMB. □

□ (b) The triangle AMB is an acute-angled triangle if the following three conditions are simultaneously satisfied: (1) $\widehat{AMB} < 90°$, (2) $\widehat{BAM} < 90°$, (3) $\widehat{ABM} < 90°$.

The required set is therefore the intersection of the following three sets: (1) the exterior of a circle with the diameter AB (see Chapter 2, Proposition **D**); (2) the half plane bounded by l_A containing the point B, with the boundary line l_A removed; (3) the half plane bounded by l_B containing the point A with the boundary line l_B removed.

The intersection is the strip between the lines l_A and l_B from which the circle with diameter AB is removed. □

□ (c) Note that every point M of the plane (not lying on the straight line AB) satisfies one of the following three conditions: either (a) $\triangle AMB$ is a right-angled triangle, or (b) $\triangle AMB$ is an acute-angled triangle or (c) $\triangle AMB$ is an obtuse-angled triangle. Note, moreover, that these conditions are, however, mutually exclusive. Hence, all the points of the plane that belong neither to (a) nor to (b) must belong to the set (c). This set is the union of a disc—that is, the region inside the circle but not including it—and two half planes (with the line AB removed). □

3.12. In a plane, two points A and B are given. Find the set of points M such that:

(a) the triangle AMB is an isosceles triangle,

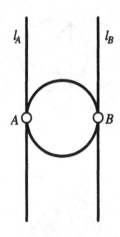

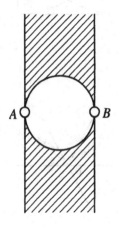

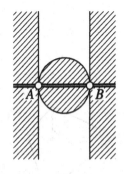

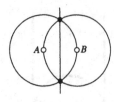

(b) the side AB is the largest side of the triangle AMB,

(c) the side AM is the largest side of the triangle AMB.

3.13. A square with sides of unit length is given in a plane. Prove that if a point of the plane lies at a distance of not more than 1 from each of the vertices of this square, then it lies at a distance of not less than 1/8 from each side of the square.

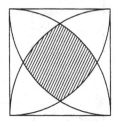

□ The set of points M at a distance of not more than 1 from each of the four vertices of the square is the intersection of four circles of unit radius, each with center at one of the vertices of the square. It is a region bounded by four arcs, and it lies in the interior of the square. This region has four "vertices"; each vertex lies at a distance of $1 - \frac{\sqrt{3}}{2}$ from the nearest side. Let us check that this number is greater than 1/8:

$$1 - \frac{\sqrt{3}}{2} > \frac{1}{8} \iff \frac{7}{8} > \frac{\sqrt{3}}{2} \iff \frac{49}{16} > 3.$$

It is thus clear that all the points of our set are at a distance of more than 1/8 from the sides of the square. □

3.14. Three straight lines passing through a point O of the plane divide the plane into six congruent angles. Prove that if the distance of the point M from each of the straight lines is less than 1, then the distance $|OM|$ is less than 7/6.

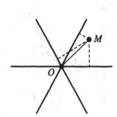

3.15. Given a square $ABCD$, find the set of points that are closer to the straight line AB than to the lines BC, CD and DA.

3.16. Given a triangle ABC, find the set of points M in the plane for which the area of each of the triangles AMB, BMC, and CMA is less than that of the triangle ABC.

3.17. Circles are drawn with the sides of an arbitrary convex quadrilateral $ABCD$ as diameters. Prove that they cover the whole quadrilateral.

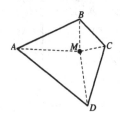

□ Assume that inside the quadrilateral there exists a point M lying outside the circles. Then according to Chapter 2, Proposition E, all the angles AMB, BMC,

42

CMD and DMA are acute and their sum is less than $360°$, which is impossible. □

3.18*. A portion of a forest has the form of a convex polygon of area A and perimeter P. Prove that we can find a point in the forest that is at a distance greater than A/P from the edge of the forest.

3.19*. A square $ABCD$ is given in a plane. Find the set of points M such that $\overset{\frown}{AMB} = \overset{\frown}{CMD}$.

In the problems that follow, we have to deal with the union of an infinite number of sets.

3.20. (a) A point O is given. Consider the family of circles of radius 3 whose centers are located at a distance of 5 units from the point O, and the family of circles of radius 5 whose centers are located at a distance of 3 units from the point O. Prove that the union of the first family of circles coincides with the union of the second one.

(b) Find the set of midpoints of the segments which have one end lying on one given circle and the other end on the other given circle.

□ (b) Denote the radii of the given circles by r_1 and r_2 and their centers by O_1 and O_2, respectively. Let us first fix some point K of the first circle and find the set of midpoints of the segments that have one end at the point K. This set will obviously be a circle of radius $r_2/2$ with its center Q at the midpoint of the segment KO_2. (This circle is the result of the similarity transformation of the circle (O_2, r_2) with coefficient 1/2 and center K.) Note that the point Q lies at a distance $r_1/2$ from the point P, the midpoint of the segment O_1O_2.

If we move the point K around the circle (O_1, r_1), then the point Q will move around the circle of radius $r_1/2$, with center at the point P. Thus, the required set is the union of all circles of radius $r_2/2$ that have their centers lying on a circle of radius $r_1/2$ with its center at the point P.

What this union of an infinite number of circles turns out to be can be seen in the figure.

Consequently, the set of all points satisfying the condition given in the problem is a ring with external radius $(r_1+r_2)/2$ and internal radius $|r_1-r_2|/2$. When $r_1 = r_2$ this set becomes a circle. □

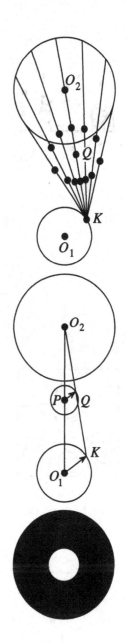

3.21. A point O is located on a straight line l, the boundary line of a half plane. In this half plane, n vectors of unit length are drawn from the point O. Prove that if n is odd, the length of the sum of these vectors is not less than 1. ↓

3.22. A straight road passes through a village A surrounded by meadows on all sides. A man can walk at a speed of 5 km/h along the road and at 2 km/h through the meadows. If he starts from A and walks for one hour, what is the set of all possible points he can reach?

The "cheese" problem

3.23. Is it always possible to cut a square piece of cheese with cavities into convex pieces so that there is only a single cavity in each piece?

Formulated mathematically, this problem is as follows.

Several pairwise nonintersecting circles are located inside a square. Is it possible to divide this square into convex polygons such that in each of them there is exactly one circle? (Recall that a polygon in *convex* if, given any two points in the polygon, the line connecting them lies entirely within the polygon.)

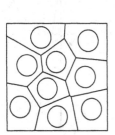

□ The answer turns out to be affirmative. In any particular example in which the number of circles is not large, one can easily divide the square into convex polygons. But to give a general proof, we must give a method for partitioning the square which can be used for any number and any positioning of the circles.

Let us first consider a simpler problem: *the radii of all the circles will be taken to be equal.* We propose the following method for partitioning the square. We shall first describe it briefly in a single sentence:

Adjoin to each of the circles those points of the square that are nearer to this circle than to the other circles; these sets will be the required convex polygons ⟨?⟩.

We shall explain this in more detail. Denote the centers of the given circles by $C_1, C_2, \ldots, C_n$. Let C_i be one of these centers. Let us find the set of points whose distance from C_i is not greater than the distance from the other centers C_j. The set of points of the plane that are nearer to C_i than to C_j (for a fixed j) is a half plane bounded by the perpendicular bisector of the segment C_iC_j (see **A**). We are interested in the points that are nearer to C_i than to the other centers; i.e., the points belonging to all such half planes corresponding to the different C_j ($j \neq i$). This set of points, which is the intersection of all these $(n-1)$ half planes, will clearly be a convex polygon. $\langle ? \rangle$ Since each half plane contains the point C_i and the entire circle with its center at C_i (the circles with centers C_i and C_j do not intersect · and have equal radii), the intersection also contains the circle with its center at C_i. There is such a polygon

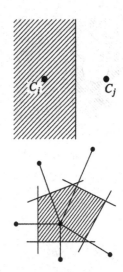

$$\{M : |MC_i| \leq |MC_j| \text{ for all } j \neq i\}$$

for every center C_i. It is clear that these polygons cover the entire square and have no interior points in common. In order to determine to which particular polygon the point M belongs, it is sufficient to answer the question, "Which of the centers C_i is closest to the point M?" If there are two or more such centers "closest to M," then M lies on one of the perpendicular bisectors—that is, on a boundary line or line of partition between the polygons. Thus, the square is divided into convex polygons each of which contains exactly one circle.

As a good example, let us consider the case when *the centers of the circles are located at the nodes of a net formed by similar parallelograms.*

Our method of partition can be simply described in the following way.

Draw the minor diagonals in all the parallelograms of the net. This will yield a net with the same nodes, made from similar acute-angled triangles. Inside each triangle, draw the midperpendiculars. The hexagons thus obtained form the required partition of the square. Thus, we have analyzed the case in Problem **3.23** when all the circles have equal radii.

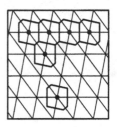

In the general case, when the *radii of the circles are different*, the square can be divided in the following manner. From each point located outside the given circles, draw tangents to all the circles. The set corresponding to the circle γ will consist of the points of the circle γ and those points for which the lengths of

45

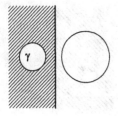

the tangent to the circle γ is less than the length of the tangents to the remaining circles. This set is the intersection of several half planes containing the circle γ. The boundary lines of these half planes will be the radical axes of the circle γ and each of the other circles (see problems **2.9** and **3.5**). In this way the whole square will be represented as the union of convex polygons, with no interior points in common, such that each polygon contains its own circle. $\square$

CHAPTER 4

Maximum and Minimum

This chapter starts with very simple exercises concerning the greatest and the least possible values of a certain quantity. The chapter ends, however, with complicated research questions. Maxima and minima problems can usually be reduced to the examination of some function which is given analytically, but here we have a collection of problems in which geometric considerations prove to be more effective. We will see how, in the solution of similar problems, different sets of points are used.

4.1. At what angle to the bank of a river should one direct a boat so that, while crossing the river, it is carried as little as possible by the current, assuming that the speed of the current is 6 km/h and the speed of the boat in still water is 3 km/h?

□ *Answer*: at an angle of 60°. We have to direct the boat so that its absolute velocity (the velocity relative to the bank) makes the largest possible angle with the bank ⟨?⟩ (see the figure). Let the vector $\overrightarrow{OA}$ be the velocity of the river current and $\overrightarrow{AM}$ the velocity of the boat relative to the water. The sum $\overrightarrow{OA} + \overrightarrow{AM} = \overrightarrow{OM}$ represents the absolute velocity of the boat. The length of the vector $\overrightarrow{AM}$ is equal to 3 and we can direct this vector arbitrarily. The set of possible positions of the point M is a circle of radius 3, with center at the point A. It is clear that among all the vectors $\overrightarrow{OM}$, only $\overrightarrow{OM_0}$, which is directed along the tangent to the circle, makes the largest angle with the bank.

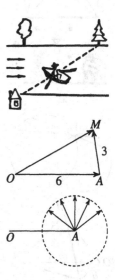

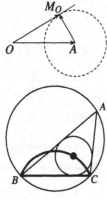

We obtain a right-angled triangle, one leg of which is equal to half the hypotenuse. Such a triangle has one angle equal to 60°. □

4.2. From the triangles with given base BC and $\hat{A} = \varphi$, select the one whose inscribed circle has the largest radius.

□ Let us consider the points A which lie on one side of the straight line BC and for which $\widehat{BAC} = \varphi$. The set of centers of the inscribed circles of the triangle ABC is the arc of a circle with endpoints B and C (see 3.7b). It is obvious that the isosceles triangle will have the largest radius of the inscribed circle. □

4.3. From all the triangles with a given base and a given vertex angle, select the triangle with the largest area.

4.4. Two pedestrians walk along two mutually perpendicular roads, one at a speed of u and the other at a speed of v. When the first pedestrian crosses the second pedestrian's road, the second pedestrian still has d kilometers to go to reach the crossing. What will be the minimum distance between them? ↓

4.5. A straight road passes through a village A surrounded by meadows on all sides. A man can walk at a speed of 5 km/h along the road and at 2 km/h through the meadows (in any direction).

Along what route should the man walk to go as quickly as possible from village A to cottage B, which is situated at a distance of 13 km from the village and at a distance of 5 km from the road?

4.6. Two intersecting circles are given. Draw a straight line l through one of their points of intersection, A, such that the distance between l and the other points of intersection (that is, the points other than A where the circles intersect) is as large as possible. ↓

4.7. A point O is given in a plane. One of the vertices of an equilateral triangle must lie at a distance a from the point O and a second vertex at a distance b. What is the maximum distance from O at which the third vertex can be situated?

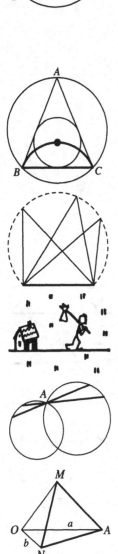

48

□ *Answer: a + b.* Let AMN be an equilateral triangle for which $|OA| = a$ and $|ON| = b$. In order to answer the question we may restrict ourselves to triangles having a vertex fixed at a definite point A; for, when the triangle is rotated as a rigid body about the point O, none of the distances changes. Thus, we consider the point A fixed at a distance a from O, while N runs around the circle of radius b with center O. What position may the point M occupy? The answer has already been obtained in Problem **1.9**: M lies on the circle obtained from the given circle by rotating it through 60° about the point A.[1] The center O' of the rotated circle obviously lies at a distance a from the point O (since $\triangle OO'A$ is equilateral). The radius of the rotated circle, as for the given one, is equal to b. Therefore, the maximum distance from O to the third vertex M is equal to $a + b$. □

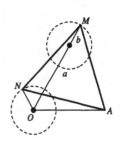

From this problem, we can deduce the following interesting corollary: the distance from an arbitrary point in the plane to one of the vertices of an equilateral triangle is not greater than the sum of the distances from the point to the other two vertices.

4.8. What is the maximum distance at which the vertex M of a square $AKMN$ may lie from the point O, it if is known that

(a) $|OA| = |ON| = 1$;

(b) $|OA| = a, |ON| = b$?

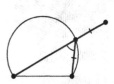

4.9. From all the triangles with a given base and a given vertex angle, select the one having the largest perimeter. ↓

Where to put the point?

4.10. A cat knows the three exits A, B, C of a mouse's hole. Where should the cat sit in order to minimize its distance to the furthest hole?

□ Let us consider circles of equal radius r with their centers at the points A, B and C. The required point

---

[1]We may take any of the circles, obtained by clockwise or counterclockwise rotation—they will lie at the same distance from O.

K—the position of our cat—is determined as follows. We must find the minimum radius r_0 for which the three regions *within* the circles overlap—that is, the minimum radius for which the three regions within the circles share a common point. This is the required point K. For if M is any other point, then it lies outside one of the circles and hence its distance from one of the vertices is greater than r_0.

In the case of an acute-angled triangle ABC, the point K is the center of the circumscribed circle, and in the case of a right-angled or an obtuse-angled triangle ABC, the point K is the midpoint of the largest side. $\square$

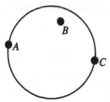

$\square$ The point K can also be found in the following way ⟨?⟩. Consider the circle of minimum radius which surrounds all three points. Then the point K is its center. $\square$

We shall give another approach to the solution of Problem **4.10**.

$\square$ Divide the plane into three sets of points:

(a) $\{M: |MA| \geq |MB| \text{ and } |MA| \geq |MC|\}$,

(b) $\{M: |MB| \geq |MA| \text{ and } |MB| \geq |MC|\}$,

(c) $\{M: |MC| \geq |MB| \text{ and } |MC| \geq |MA|\}$.

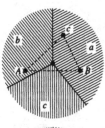

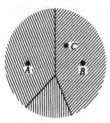

These three regions in the plane are bounded by the perpendicular bisectors of the sides of the triangle ABC. If the cat M sits in region (a), then the vertex farthest from it will be A; if it sits in region (b), then the vertex farthest from it is B; if the cat sits in region (c), then the vertex farthest from it is C.

If ABC is an acute-angled triangle, then in each of the three cases the best thing for the cat to do is to sit at the "vertex" of the corresponding region ((a), (b) or (c)): that is, the cat should sit at the center of the circumcircle.

If ABC is a right-angled or an obtuse-angled triangle, then obviously the best thing for the cat to do is to sit at the midpoint of the largest side of the triangle. $\square$

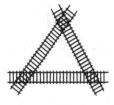

4.11. A bear lives in a part of a forest surrounded by three straight railway lines. At which point of the forest should he build his den in order to maximize his distance from the nearest railway line?

4.12*. (a) Three crocodiles live in a circular lake. Where should they lie so that the maximum distance from any point of the lake to the nearest crocodile is as small as possible?

(b) Solve the same problem when there are four crocodiles.

The motorboat problem

4.13*. A searchlight is located on a small island. Its beam lights up the sea's surface along a distance $a = 1$ km. The searchlight rotates uniformly about a vertical axis at a speed of one revolution in the time interval $T = 1$ min. A motorboat which moves at speed v must reach the island without being caught by the searchlight beam. What is the minimum value of v for which this is possible?

☐ Let us call the circle of radius a which is illuminated by the searchlight beam the "detection circle." It is clear that for the motorboat the best thing to do is to enter this circle at a point A through which the beam of the searchlight has just passed.

If the motorboat heads straight for the island, it will reach the island in time a/v. In order to guarantee the beam of the searchlight not catch the motorboat in this time, it is essential that the beam not complete a full revolution within this time, i.e., that the inequality $a/v < T$ hold, from which

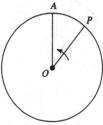

Detection circle

$$v > a/T = 60 \text{ km/h}.$$

Thus, we have shown that the motorboat may reach the island unnoticed when $v > 60$ km/h. But, of course, it does not follow that 60 km/h is the minimum value of the speed of the boat for which this is possible— namely, that moving along the segment AO is the best possible course which the captain of the motorboat can select. Indeed, as we shall see, this is not the case at all.[2]

---

[2]Before reading the solution further, try to guess a route for the motorboat to reach the island with a smaller value of v.

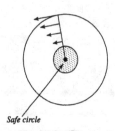

Safe circle

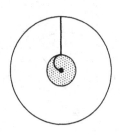

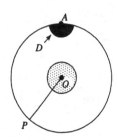

Note that the linear velocity of the beam OP of the searchlight is different at different points: the nearer the point is to the center, the smaller its velocity. The angular velocity of the beam is equal to $2\pi/T$. The motorboat can easily travel ahead of the beam, around a circle of radius $r = vT/2\pi$, since the velocity of the boat here is equal to the linear velocity of the corresponding point of the beam. Outside the circle of radius r, with center O, the speed of the beam is greater, and inside this circle (we shall call it the "safe circle") the speed of the beam is less than v.

If the motorboat is able to reach some point of the safe circle without hindrance, then it can clearly reach the island unnoticed.

For example, one of the possible courses inside the safe circle is to traverse a circle of radius $r/2$. If the motorboat K moves around this circle with a speed v, then the segment KO will rotate about O with the same angular velocity with which the boat would have moved around a circle of radius r, i.e., with the same angular velocity as the beam of the searchlight (see Problem **0.3**). Hence the boat will not be caught by the beam.

Thus, the aim of the motorboat is simply to reach the safe circle!

If the motorboat heads straight to the search light along the radius AO, then it will be able to reach the safe circle without being detected by the beam of the searchlight if

$$v > \frac{1}{1 + (1/2\pi)} \frac{a}{T} \cong 0.862 \frac{a}{T} = 51.7 \text{ km/h}.$$

We have been able to improve our previous estimate of the minimum speed of the motorboat. But we shall see that even this is not the best possible value!

Now let us find the *minimum* value of the speed v for which the motorboat can reach the island unnoticed.

The set of points in the detection circle which the motorboat can reach in time t is the region bounded by an arc of radius vt with center at the point A. Of these points the motorboat may reach, unnoticed, those which are located to the left of the beam OP.

52

Denote the set of these "reachable' points by D. The diagrams show how this set changes with time until the moment when one of the following possibilities occurs:

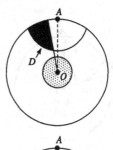

(1) If the speed v is not sufficiently high, then at some instant t, the set D will be totally exhausted without the safe circle being reached: this means that in the time t, the boat will be spotted, i.e., for this speed value the motorboat will not be able to reach the island. Note that at the last moment $t = t_0$ the beam OP will touch the arc of radius vt_0 with center A at some point L. Clearly the point L is located outside the safe circle (otherwise the motorboat would be able to reach the island). Moreover, as the speed v increases, the detection time t_0 also increases, and the distance from the point L to the island decreases.

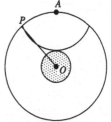

(2) If the speed v is greater than some value v_0, then the set D extends to the safe circle at some point of time. This means that the motorboat can reach the island when $v > v_0$.

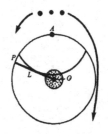

The minimum value of the speed v_0 corresponds to the case when the beam OP touches the arc of radius vt_0 right on the circumference of the safe circle. To find the value v_0, denote the value of the angle NOA by β and use the following inequalities:

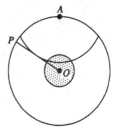

$$|NO| = r = \frac{v_0 T}{2\pi}, \quad |AN| = v_0 t_0,$$

$$\frac{|AN|}{|NO|} = \tan\beta, \quad \frac{2\pi + \beta}{t_0} = \frac{2\pi}{T},$$

$$|NO| = a\cos\beta.$$

From the first and last equation we find that

$$v_0 = (2\pi a \cos\beta)/T,$$

and from the first four equations we obtain an equation for β:

$$2\pi + \beta = \tan\beta.$$

We can only solve this equation approximately—for instance, with the help of a computer. The value of β turns out to be approximately $0.92\pi/2$, and hence

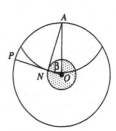

$$v_0 \cong 0.8a/T \cong 48 \text{ km/h.}$$

53

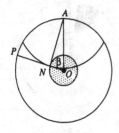

When the speed is greater than v_0, the motorboat is able to reach the safe circle. $\square$

4.14*. (a) A boy is swimming in the middle of a circular swimming pool. His father, who is standing at the edge of the swimming pool, does not know how to swim, but can run four times faster than his son can swim. The boy can run faster than his father. The boy wants to run away. Is it possible for him to do so?

(b) At what ratio between the speeds v and u (v is the speed at which the boy swims, u is the speed at which his father runs) will the boy be unable to run away?

CHAPTER 5

Level Curves

In this chapter we discuss the problems and theorems of the previous chapter using some new terminology. The concepts we are going to examine here revolve around *functions defined on a plane* and their *level curves*. These are especially useful in the solutions to problems involving maxima and minima.

The bus problem

5.1. A tourist bus is travelling along a straight highway. A palace is situated by the side of the highway at some angle to the highway. At what point on the highway should the bus stop for the tourists to be able to see the façade of the palace from the bus in the best possible way?

Mathematically, the problem may be formulated as follows.

A straight line l and a segment AB which does not intersect it are given. Find the point P on the line l for which the angle APB assumes its maximum value.

Let us first have a look at how the angle AMB changes when the point M moves along the straight line l. In other words, let us look at the behavior of the function f which relates each point M on the line to the size of the corresponding angle $\widehat{AMB}$.

It is easy to draw a rough graph of this function. (Remember that we draw graphs in the following way: above each point M of our straight line, we plot a point at a distance of $f(M) = \widehat{AMB}$.)

The problem can be solved analytically: introduce coordinates on the straight line l, express the value of

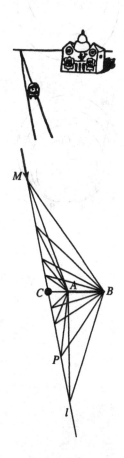

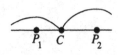

P_1 C P_2

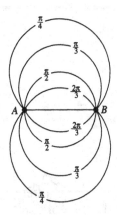

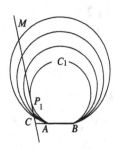

the angle AMB in terms of the x-coordinate of the point M, and find the value of x for which the resulting function assumes its maximum. However, the formula for $f(x)$ is quite complicated.

We shall give a more elementary and instructive solution. But to do this, we have to study how the value of the angle AMB depends on the position of the point M *in the whole plane* (and not only on the straight line l).

□ The set of points M in the plane for which the angle AMB assumes a given value φ is a pair of symmetric arcs with their endpoints at A and B (see Chapter 2, Proposition E). If these arcs are drawn for different values of φ (where $0 < \varphi < \pi$), we get a family of arcs that cover the whole plane except for the straight line AB. Some of these arcs are drawn in the figure, and on each arc, the corresponding value of φ is marked. For example, a circle with diameter AB corresponds to the value $\varphi = \pi/2$.

We now consider only the points M on the straight line l. From these, we have to select the specific point for which the angle AMB assumes its maximum value. Some arc from our family passes through each such point: if $\overset{\frown}{AMB} = \varphi$, the point M lies on the arc corresponding to the value φ. Thus, the problem is reduced to the following: from all the arcs crossing the line l, select the one that corresponds to the maximum value of $\overset{\frown}{AMB} = \varphi$.

We will examine the part of the straight line l located to one side of the point C, the point of intersection of the straight line AB with l. (We will not consider the case when the segment AB is parallel to the line l—we leave that to the reader). We draw the arc c_1 touching this part of the straight line and prove that the segment AB subtends the maximum angle at the point of tangency P_1. Any point M of the straight line l, except P_1, lies outside the segment cut off by the arc c_1. As we know (see Proposition E, p. 22), from this it follows that $\overset{\frown}{AMB} < \overset{\frown}{AP_1B}$.

It is obvious that on the other side of the point C everything will be exactly the same: the point P_2, at which the angle subtended by the segment AB is a

56

maximum, is also the point of tangency of the straight line with one of the arcs of our family.

We have thus proved that the required point P in our problem coincides with one of the points P_1 or P_2 at which the circles passing through the points A and B touch the straight line l.

We should select P as the point for which the angle PCA is an acute angle. If the segment AB is perpendicular to the line l, then from symmetry considerations, it is immediately obvious that the points P_1 and P_2 are completely equivalent; hence the number of solutions to the problem, in this case, is two. (However, the tourists in any case must select that point P_1 or P_2 from which the façade of the palace is visible.)

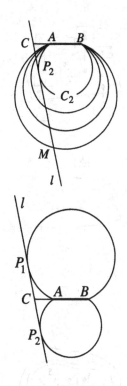

Functions on a plane

The main idea of the solution of Problem **5.1** is to consider *over the whole plane* the function f which relates each point to the corresponding angle value $\widehat{AMB}$, i.e., $f(M) = \widehat{AMB}$.

In the previous chapters we have already encountered various types of functions. Apart from the simplest functions on a plane, such as $f(M) = |OM|$, $f(M) = \rho(l, M)$, $f(M) = \widehat{ABM}$ (where O, A, B are given points and l is a given straight line), we considered the sums, the differences, and the ratios of such functions, as well as other combinations of them.

Level curves

Most of the conditions by which our sets of points were defined can be represented in the following way. On a plane (or on some region of it) we are given a function f and we need to find the set of points M for which this function assumes a given value h, i.e.,

$$\{M : f(M) = h\}.$$

As a rule, for every fixed number h, this set is some curve; thus the plane is covered by curves, called the *level curves* of the function f. In solving Problem **5.1**, we have drawn the level curves of the function $f(M) = \widehat{AMB}$.

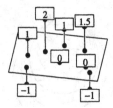

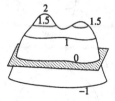

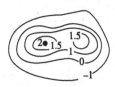

The graph of a function

Let us now explain where the term "level curves" comes from. For functions defined on a plane, we may draw graphs in exactly the same way as for functions of the form $y = f(x)$ that are defined on a straight line—except that now we have to draw the graph in space. Let us suppose that the plane on which our function f is defined is horizontal, and for each point M of this plane let us plot the point located at a distance $|f(M)|$ above the point M if $f(M) > 0$, and at a distance $|f(M)|$ below the point M if $f(M) < 0$. The points plotted in such a manner usually form some surface, which is called the *graph of the function* f. In other words, if we introduce coordinate system Oxy on the horizontal plane and direct an axis Oz vertically upwards, then the graph of the function will be the set of points with coordinates (x, y, z), where $z = f(M)$ and (x, y) are the coordinates of the point M on the plane. (If the function is not defined for all the points in the plane, but only in some region, then the graph will be located only above the points in this domain of definition.)

Hence, the level curve $\{M : f(M) = h\}$ consists of those points M above which the points of the graph are located at the same level, namely, at the height h. On pp. 60–61 we have shown the graphs of the functions whose level curves represent the sets of our alphabet.

The graph of the function $f(M) = \widehat{AMB}$ is a "mountain range" of height π above the segment AB, from which the graph gradually comes down to zero. (Remember that we constructed the graph of this function at the very beginning of the solution of Problem 5.1, but only above a particular straight line l.)

A function f of the form

$$f(M) = \lambda_1 \rho(M, l_1) + \lambda_2 \rho(M, l_2) + \cdots + \lambda_n \rho(M, l_n),$$

as mentioned in Chapter 2 (the theorem on the distances from straight lines) may be written as a linear expression

$$f(x, y) = ax + by + c$$

on each of the pieces Q into which the plane is divided by the straight lines $l_1, l_2, \ldots, l_n$.

58

Its graph will thus consist of pieces of planes, either inclined or horizontal (if $a = b = 0$). This can be seen in the examples of sets given in Propositions **C**, **I**, **J** of the "alphabet."

The level curves of such a function consist of pieces of straight lines; if the graph has a horizontal plane, then one of the level curves includes the whole piece Q of the plane.

A function f of the form

$$f(M) = \lambda_1 |MA_1|^2$$
$$+ \lambda_2 |MA_2|^2 + \cdots + \lambda_n |MA_n|^2$$

when $\lambda_1 + \lambda_2 + \cdots + \lambda_n = 0$, also reduces to a linear function on the whole plane (e.g., Proposition **F**) and in the general case, when $\lambda_1 + \lambda_2 + \cdots + \lambda_n \neq 0$ to a function of the form

$$f(M) = d |MA|^2,$$

where A is some point in the plane. Its level curves are circles (see the theorem on the squares of the distances in Chapter 2), and the graph is the surface of a *paraboloid of rotation*.

The functions $f(M) = \widehat{AMB}$ and $f(M) = |AM|/|BM|$ have perhaps the most complicated graphs in our "alphabet." Note that there is an interesting relation between the maps of the level curves of these functions: if they are drawn on a single diagram, then we get two different families of circles. However, every circle in the first family intersects every circle in the second family at a right angle. ⟨?⟩ Hence these families are said to be *orthogonal*.

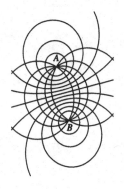

We give one more example of a simple function, one whose level curves are rays issuing from a single point and whose graph is quite a complicated surface. The function is $f(M) = \widehat{MAB}$ (where A and B are given points of a plane). Its graph above each of the half planes into which the straight line AB divides the plane is a *spiral surface*, like the surface of a screw, and it is called a *helicoid*.

59

Here the graphs of functions corresponding to the propositions of our "alphabet" are depicted and, under each one of them, there is a map of the appropriate level curves.

C. $f(M) = \rho(M, l)$. The graph is a two-sided angle; the level curves are pairs of parallel lines.

D. $f(M) = |MO|$. The graph is a cone; the level curves are concentric circles.

E. $f(M) = \widehat{AMB}$. The graph is a mountain with its peak in the form of a horizontal segment, at the ends of which there are vertical drops.

F. $f(M) = |MA|^2 - |MB|^2$. The graph is a plane; the level curves are parallel straight lines.

G. $f(M) = |MA|^2 + |MB|^2$. The graph is a paraboloid of rotation, and the level curves are concentric circles.

H. $f(M) = |MA|/|MB|$. The graph has a depression near the point A; near B, it rises to infinity. The level curves are nonintersecting circles whose centers lie on the straight line AB, each pair of which, however, has the same straight line, the perpendicular bisector of the segment AB, as its radical axis.

I. $f(M) = \rho(M, l_1)/\rho(M, l_2)$. The graph is obtained in the following manner: consider a saddle-shaped surface—the "hyperbolic paraboloid" passing through the straight line l_1 and the vertical straight line passing through the point of intersection O of l_1 and l_2. The part of this surface lying below the given plane is reflected symmetrically relative to the plane. The level curves are pairs of straight lines passing through the point O.

J. $f(M) = \rho(M, l_1) + \rho(M, l_2)$. The graph is a four-sided angle. The level curves are rectangles with their diagonals belonging to l_1 and l_2.

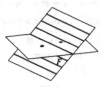

Fig. 1

60

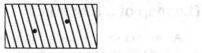

Fig. 2

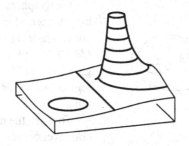

Fig. 3

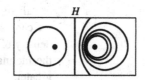

Fig. 4

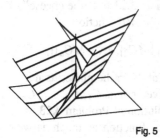

Fig. 5

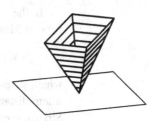

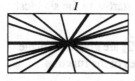

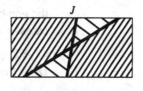

Fig. 6

61

The map of a function

As we can see, for many functions, it is difficult to draw their graphs in three-dimensional space. It is easier, as a rule, to visualize the behavior of the function on a plane by drawing the map of its level curves.

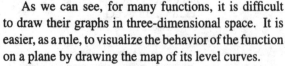

Geographers draw physical maps in the following manner. Let $f(M)$ be the height of the surface above sea-level at the point M. Then the level curves $\{M: 0 < f(M) < 200 \text{ m}\}$ are colored green, the region $\{M: f(M) > 200 \text{ m}\}$ is colored brown, and the region $\{M: f(M) < 0\}$ is colored various shades of blue.

To make the map of a function, one must draw several level curves—as many as are needed to be able to judge from them where the other curves are. Then one must mark each of them with the value of the function to which it corresponds (i.e., the value of h).

If we decide to depict the level curves at equal intervals of the functional values $0, \pm d, \pm 2d, \ldots$, then we can estimate the inclination of the graph from the density of the level curves: where there are more lines the inclination of the graph to the horizontal plane is greater.

Boundary lines

In the solution of Problem **3.23** (on "the cheese") we considered a quite complicated function

$$f(M) = \min\{|MC_1|, |MC_2|, \ldots, |MC_n|\},$$

which gives, for every point M in the plane, its minimum distance from the given points $C_1, C_2, \ldots, C_n$. Strictly speaking, in the solution of Problem **3.23**, we did not need this particular function as much as we needed the boundary lines associated to it; those boundary lines partitioned the plane into polygonal regions. Let us try to visualize the map of the level curves and the graph of this function. We start with the simplest cases: $n = 2$ and $n = 3$.

5.2. (a) Two points C_1 and C_2 are given in a plane. Draw the map of the level curves of the function

$$f(M) = \min\{|MC_1|, |MC_2|\}.$$

(b) Three points C_1, C_2, C_3 are given in a plane. Draw the map of the level curves of the function $f(M) = \min\{|MC_1|, |MC_2|, |MC_3|\}$.

□ (a) Consider the set of points M for which $|MC_1| = |MC_2|$. As we know, this set of points forms the perpendicular bisector of the segment C_1C_2. This perpendicular bisector divides the plane into two half planes. The points of one half plane are closer to C_1, and the points of the other are closer to C_2.

Thus, in one half plane $f(M) = |MC_1|$, and in the other $f(M) = |MC_2|$. Hence, in the first half plane, we must draw the curves of the function $f(M) = |MC_1|$—these are circles—and then reflect this map symmetrically through the perpendicular bisector.

(b) Consider the sets of points where $|MC_1| = |MC_2|$, where $|MC_2| = |MC_3|$ and where $|MC_1| = |MC_3|$. We looked at them in Problem **3.1**. They are the three midperpendiculars of the triangle $C_1C_2C_3$, which intersect at a single point O. These three rays, formed by the midperpendiculars with their initial point at O, partition the plane into three regions. Clearly in the region containing the point C_1, $f(M) = |MC_1|$; in the region containing the point C_2, $f(M) = |MC_2|$; and in the region containing the point C_3, $f(M) = |MC_3|$. Thus, the map of the function $f(M) = \min\{|MC_1|, |MC_2|, |MC_3|\}$ is the union of three maps, joined along the lines of partition, i.e., along the three rays. □

The graph of the function

$$f(M) = \min\{|MC_1|, |MC_2|, \ldots, |MC_n|\}$$

may be visualized in the following manner. If a uniform layer of sand is placed in a box and holes are made in the bottom of the box at the points $C_1, C_2, \ldots, C_n$ through which the sand comes out, then around each hole a "funnel" is formed. The surface of all these "funnels" forms the graph of the function f. (We must, of course, use sand such that the angle of its natural slope is equal to $45°$; furthermore, we must use a sufficiently thick layer of it.)

Let us now return to problems **3.11** and **3.12**. We can find functions defined on a plane in these problems as well.

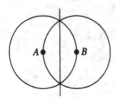

5.3. Let the points A and B be given in a plane. Draw the map of the level curves of the functions

(a) $f(M) = \max\{\widehat{AMB}, \widehat{BAM}, \widehat{MBA}\}$,

(b) $f(M) = \min\{|AM|, |MB|, |AB|\}$,

and describe their graphs.

Extrema of functions

Let f be a given function defined on a plane. Imagine its graph as a hilly area. The maximum values of $f(M)$ correspond to the heights of the "hill tops" of its graph, and the minimum values to the depths of the valleys or depressions. On the map of the level curves of a function, the hill tops and the depressions are, as a rule, circled by level curves. For instance, for the function $f(M) = |MA|^2 + |MB|^2$, the minimum point M_0 is the midpoint of the segment AB, and the level curves are concentric circles with their centers at the point M_0.

We get a more complicated picture for the function $f(M) = \widehat{AMB}$. This function assumes its maximum value π at all the points of the segment AB, and its minimum value 0, at the remaining points of the straight line AB. The transition from the maximum to the minimum value at the points A and B is not gradual (f is not defined at these points): here the graph has vertical drops.

At the beginning of the chapter we used a map of level curves for the solution of Problem **5.1**. This is also a problem of finding a maximum, but of a different type. The problem may be stated generally in the following way: *Find the maximum and minimum values assumed on some curve γ by a function defined on a plane* (in the problem we looked at, γ was a straight line). The observation we made in Problem **5.1** also holds for these similar problems: the *maximum* (*and minimum*) *values are usually assumed at the points where γ touches the level curves of the function f.*[1]

Let us assume that the maximum value of the function f on the curve γ is attained at the point P and

---

[1] Or at the point where the function f itself reaches a maximum, if the curve γ passes through such a point.

is equal to $f(P) = c$. Then the curve γ cannot enter the region $\{M: f(M) > c\}$: it must belong entirely to the complementary region $\{M: f(M) \leq c\}$. The point P lies on the line separating these regions, i.e., on the level curve $\{M: f(M) = c\}$. Thus, the curve γ cannot cross the level curve $\{M: f(M) = c\}$, i.e., it must touch this line at the point P.

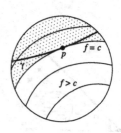

You have seen how this "tangency principle" for finding an extremum arose in the problems in Chapter 4.

In these problems we looked for the maximum or minimum of the simple functions:

$$f(M) = \rho(M, l), \quad f(M) = \widehat{MOA},$$
$$f(M) = |MA|$$

on a given curve γ. The level curves corresponding to the extreme value were touched by the curve γ. As a rule, this curve γ was a circle.

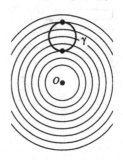

Some of the following problems also reduce to problems of finding the maximum (or minimum) of a function on a given circle or straight line.

5.4. (a) On the hypotenuse of a given right-angled triangle, find the point for which the distance between its projections onto the legs is minimized. (Recall that the *legs* of a right triangle are the sides that define the right angle.)

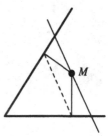

(b)* On a given straight line, find a point M such that the distance between its projections onto the sides of a given angle is minimized. ↓

5.5. Given a circle with center O and a point A inside it, find a point M on the circle for which the value of the angle AMO is maximized.

5.6. Suppose points A and B are given. On a given circle γ, find:

(a) a point M such that the sum of the squares of the distances from M to the points A and B is a minimum.

(b) A point M such that the difference between the squares of the distances from M to the points A and B is a minimum.

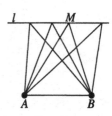

5.7. Given a straight line l and a segment AB parallel to it, find the positions of the point M on the straight

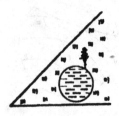

line l for which the quantity $|AM|/|MB|$ assumes its maximum or minimum value. ↓

5.8. A lake is situated between two straight roads. Where, on the edge of the lake, should a resort be built in order to minimize the sum of the distances from the resort to the two roads? Consider the cases when the lake is (a) circular, (b) rectangular.

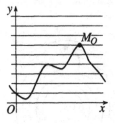

Note that in finding the maximum of a function of a single variable, $y = f(x)$, we are guided by the "tangency principle." Suppose we draw the graph of the function f on a plane. This graph will be some kind of curve. To find the maximum value of the function f, we must find the highest point on the graph. It is clear that to do this, we must draw a straight line that is both parallel to the axis Ox and tangent to the graph. Moreover, this tangent line should be drawn in such a way that the entire graph lies below it.

CHAPTER 6

Quadratic Curves

Ellipses, hyperbolas, parabolas

So far, we have limited ourselves to the lines which are thoroughly studied at school: namely, straight lines and circles. All the propositions of our "alphabet" from **A** to **J** involved only these. In this chapter we are going to learn about some other curves: ellipses, hyperbolas and parabolas. Taken together, these curves are called *conic sections* or simply *conics*, since they may all be obtained as the intersection of a plane with the surface of a cone, as is shown in the figure on pp. 76–77.

We first define ellipses, hyperbolas, and parabolas geometrically, as a continuation of our "alphabet" from Chapter 2. They will appear later as envelopes of families of lines. Finally, using analytic geometry, we will find that these curves may be defined by second-order algebraic equations. The proof of the equivalence of these definitions is not simple. However, they are all useful, since each new definition allows us to solve a new class of problems with greater ease.

Thus, let us continue our "alphabet" with the new propositions, **K, L, M**, and, a little later, **N**.

K. *The Ellipse.* Let A and B be two given points. Let us consider the *set of points M in the plane, the sum of whose distances from A and B is equal to a constant.*

Following standard convention, we denote this constant by $2a$, and we denote the distance $|AB|$ between the points A and B by $2c$. Note that when $a \leq c$, this set is of little interest: if $a < c$, then the required set is empty, as there is not a single point M on the plane for

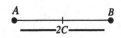

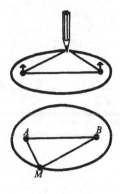

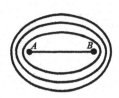

which $|AM| + |MB| < |AB|$; when $a = c$, the set is simply the segment AB.

To see what happens when $a > c$, proceed as follows. Fix two nails at A and B, put a loop of thread of length $2(a + c)$ around them, stretch the thread taut with a pencil, and draw a curve with the pencil, all the while continuing to keep the thread taut. You will get a closed curve. This curve is called an "ellipse." The points A and B are called the *foci* of the ellipse. From the definition of an ellipse, it is clear that it has two axes of symmetry: the straight line AB and its perpendicular bisector, which passes through the midpoint O of AB. The segments of these straight lines which lie inside the ellipse are called its *axes*, and the point O is called the *center* of the ellipse.

By altering the length of the thread, we can draw a whole family of ellipses with the same foci; in other words, we can draw the map of the level curves of the function

$$f(M) = |MA| + |MB|.$$

L. *The Hyperbola.* Let A and B be two given points. Consider the *set of points, the difference of whose distances from A and B is equal in absolute value to a constant* $2a$ $(a > 0)$.

Let $|AB| = 2c$ as before. If $a > c$, then the set **L** is empty, as there is not a single point M in the plane for which $|AM| - |MB| > |AB|$ or $|MB| - |AM| > |AB|$. When $a = c$, the set **L** is a pair of rays of the straight line AB—we must exclude the line *segment AB* from the entire straight line AB.

In the case when $a < c$, the set **L** consists of the two lines (branches) shown in the figure (one is the set $\{M : |MA| - |MB| = 2a\}$ and the other—$\{M : |MB| - |MA| = 2a\}$). This set is called a *hyperbola* and the points A and B are called its *foci*.

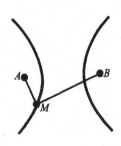

From the definition of the set **L**, it is clear that a hyperbola has two axes of symmetry. The midpoint of the segment AB is called the *center* of the hyperbola.

In order to get the whole map of the level curves of the function

$$f(M) = ||MA| - |MB||$$

68

we should also include the perpendicular bisector of the
segment AB (it corresponds to the value $f(M) = 0$)
in the family of hyperbolas with foci at A and B.

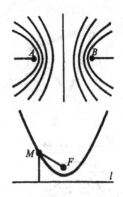

M. *The Parabola. A parabola is the set of points M
equidistant from a given point F and a given straight
line l.*

The point F is called the *focus* of the parabola, and
the straight line l is called its *directrix*. The parabola
has a single axis of symmetry, which passes through
the focus F and is perpendicular to the directrix.

Let us summarize our initial results. We have added
the following sets to our "alphabet":

K. $\{M : |MA| + |MB| = 2a\}$,

L. $\{M : ||MA| - |MB|| = 2a\}$,

M. $\{M : |MF| = \rho(M, l)\}$.

Now we know that if a problem reduces to one of
the sets **M**, **K**, or **L**, then the answer will be a parabola,
an ellipse or a hyperbola, respectively. Of course, in
the answer, we should indicate not only the name of
the curve but also its dimensions and its position, for
instance, by giving the foci and the number a.

6.1. The points A and B are given in a plane. Find
the set of points M for which:

(a) the perimeter of the triangle AMB is equal to a
constant p,

(b) the perimeter of the triangle AMB is not greater
than p,

(c) the difference $|MA| - |MB|$ is not less than d.

6.2. Suppose we are given a segment AB and a
point T lying on it. Find the set of points M for which
the circle inscribed in the triangle AMB is tangent to
the side AB at the point T.

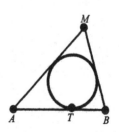

6.3. Find the set of centers of the circles in the
following cases. The circles are tangent to:

(a) a given straight line and pass through a given
point;

(b) a given circle and pass through a given point
inside the circle;

(c) a given circle and pass through a given point
outside the circle;

(d) a given circle and a given straight line;

(e)* two given circles. ↓

6.4. On a hinged closed polygon $ABCD$, for which $|AD| = |BC| = a$ and $|AB| = |CD| = b$, the link AD is fixed.

Find the set of points of intersection of the straight lines AB and CD,

(a) if $a < b$;

(b) if $a > b$.

6.5. (a) Two points A and B are given in a plane. The distance between them is an integer n (in the figure $n = 12$). Suppose we draw all the circles with integer-valued radii centered at either A or B. On the net of points thus obtained, a sequence of nodes (the points of intersection of the circles) is marked, in which any two neighboring nodes are opposite vertices of a curvilinear quadrilateral. Prove that all the points of the sequence lie either on an ellipse or on a hyperbola.

(b) Suppose we are given a straight line l in the plane and a point F lying on l. Suppose we draw all the circles of integer-valued radii with center F, and all the straight lines which are parallel to l and which lie at some integer-valued distance from l. Prove that all the points of the sequence of nodes on the resulting net, constructed as in part (a), lie on a parabola with focus F.

If we rotate a parabola, an ellipse, or a hyperbola in space about its axes of symmetry, we will obtain certain surfaces. These surfaces are called, respectively, a *paraboloid of rotation*, an *ellipsoid of rotation*, or a *hyperboloid of rotation*.

Foci and tangents

Many interesting problems concerning ellipses, hyperbolas and parabolas are connected with the properties of the tangents to these curves. We shall obtain the basic property of tangents to the ellipse by comparing two solutions of the following simple construction problem.

6.6. Suppose we are given a straight line l and two points A and B, one on each side of l. Given any point

X on l, consider the sum of the distances from X to the points A and B. At what point is this sum—namely, $|AX| + |XB|$—minimized?

☐ Consider the point A' symmetric to the point A relative to the straight line l. For any point M on this straight line, $|A'M| = |AM|$. Hence the sum $|AM| + |MB| = |A'M| + |MB|$ assumes its minimum value $|A'B|$ at the point of intersection X of the segment $A'B$ with the line l. ☐

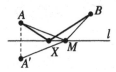

Note that the point X has the property that the *segments AX and BX make equal angles with the straight line l.*

If we had solved Problem **6.6** by the general scheme described in Chapter 5 using level curves, we would have proceeded as follows. Construct the family of ellipses corresponding to the parameter c with foci at A and B, $\{M: |AM| + |MB| = c\}$, and select from this family the particular ellipse that is tangent to the straight line l.

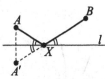

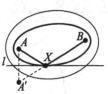

Thus, the *point X is a point of tangency of an ellipse* (with foci at A and B) *and the straight line l.* All other points M on the straight line apart from X are located outside the ellipse, i.e., for these points the sum $|AM| + |MB|$ is greater than c.

Comparing the first solution with the second, we get the so-called *focal property of an ellipse: the segments connecting the point X on an ellipse with its foci make angles of equal value with the tangent drawn to the ellipse at the point X.*

This property has an immediate physical interpretation. If the surface of a reflector (for example, a headlight) has the form of a portion of an ellipsoid, and if the lamp, taken to be a point source of light, is placed at one focus A, then after reflection the rays will converge at the other focus B (the word "focus" is the Latin word for "hearth.").

The focal property of the hyperbola is completely analogous to that of the ellipse: the *segments connecting the point X of a hyperbola with its foci make angles of equal value with the tangent at the point X.* One can prove this property by solving the following problem in two different ways.

71

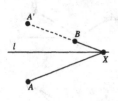

6.7. Suppose we are given a straight line l and two points A and B on opposite sides of it; the point A, however, is located at a greater distance from l than the point B. Find the point X on the straight line for which the difference between the distances $|AX| - |BX|$ is at a maximum.

One solution leads to the following answer: let A' denote the point symmetric to the point A relative to the straight line l. Then the required point X will be the point of intersection of the straight line $A'B$ with l ⟨?⟩. It is clear that for this point X, the segments AX and XB make angles of equal size with the straight line l.

The other solution (obtained by the general scheme given in Chapter 5) leads to the same answer as well: X is a point of tangency of the straight line l with a hyperbola whose foci lie at A and B. Comparing these two answers, we arrive at the focal property of a hyperbola.

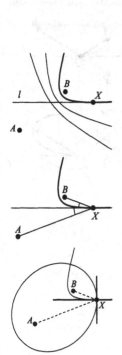

From the focal properties we can deduce another interesting property related to the families of all ellipses and hyperbolas with foci at A and B.

Consider an ellipse and a hyperbola passing through some point X. Through the point X, draw straight lines which make equal angles with the straight lines AX and BX. These straight lines are obviously perpendicular to each other.

From the focal properties, it follows that one of the straight lines is a tangent to the ellipse; the other, a tangent to the hyperbola. Thus, the tangents to the ellipse and the hyperbola are perpendicular to one another. Hence the families of ellipses and hyperbolas with foci A and B form two mutually orthogonal families; at every intersection point of a curve from one family and a curve from another family, the tangents to the two curves are perpendicular.

These two families can be clearly seen in the figure corresponding to Problem **6.5a** if the "squares" are colored, alternately, as on a chessboard.

The focal property of a parabola

Suppose a parabola has focus F and directrix l, and suppose X is some point on it. Then the *straight*

72

*line XF and the perpendicular dropped from X onto l
make equal angles with the tangent to the parabola at
the point X.*

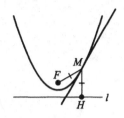

Let us prove this. Suppose H is the foot of the
perpendicular dropped from X onto l. By the defini-
tion of a parabola, we have $|XF| = |XH|$. Therefore,
the point X lies on the perpendicular bisector m of the
segment FH.

We will prove that the straight line m is a tangent to
the parabola. To do this, we will show that it has only a
single point in common with the parabola (namely the
point X), and that the entire parabola is located on one
side of m. The line m divides the plane into two half
planes. One of them consists of the points M which
are closer to F than to H.

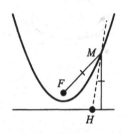

We shall show that the parabola is located in this
particular half plane, i.e., for any point M of the
parabola (except the point X) $|MF| < |MH|$. This is
immediate, as $|MF| = \rho(M, l)$ and $\rho(M, l) < |MH|$
(the perpendicular is shorter than an oblique line).

Note. For all the curves we have encountered so
far, the tangent is defined as follows: the tangent to the
curve γ at the point M_0 is the straight line l passing
through M_0 with the property that the curve γ (or at
least a part of the curve contained in some circle with
its center at M_0) lies on one side of the straight line l.

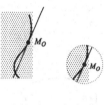

The focal property of a parabola may be used in the
following manner. If a reflector is made in the form of
a paraboloid and a light source is placed at the focus F,
then we have a projector: all the reflected rays will be
parallel to the axis of the paraboloid.

6.8. Consider all the parabolas with a given focus and a given vertical axis. They naturally fall into
two families: the parabolas of one family have their
branches extending upward, and those of the other have
their branches extending downward. Prove that any
parabola of one family is orthogonal to any parabola of
the other family; that is, prove that at the point of intersection of a curve from one family and a curve from
another family, the tangents to the curves are perpendicular.

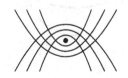

As before, these two families of parabolas can be seen clearly if the "squares" in the figure of Problem **6.5b** are colored, alternately, as on a chessboard.

The solutions to the following problems depend only on the definitions of the curves we have discussed and their corresponding focal properties.

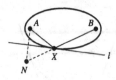

6.9. (a) Suppose we are given an ellipse with foci at A and B. Prove that the set of points symmetric to the focus A relative to any one of the tangents to the ellipse is a circle.

(b) Prove that the set formed by the feet of the perpendiculars dropped from the focus A onto the tangents to the ellipse is a circle.

□ (a) Let l be a tangent to the ellipse at the point X and let N be a point symmetric to the focus A relative to l. Then, as we know (see Problem **6.6**), the point X lies on a straight line NB and the distance

$$|NB| = |AX| + |XB|$$

is constant. Denote this distance, as before, by $2a$. Thus, the distance between N and B is constant and the required set is a circle with center at B and radius $2a$.

(b) Let M be the foot of the perpendicular dropped from the point A onto l. Clearly,

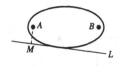

$$|AM| = \tfrac{1}{2}|AN|.$$

We know from Problem **6.9**(a) that the set of points N is a circle, so the problem reduces to the following problem. Suppose we are given a circle of radius $2a$ with center at B, and a point A inside it. Find the set of midpoints of the segments AN, where N is an arbitrary point of the circle. This set is a circle of radius a with its center at the midpoint O of the segment AB. □

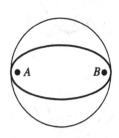

6.10. (a), (b). Prove statements (a) and (b) of Problem **6.9** for a hyperbola.

6.11. Given a parabola with focus F and directrix l:

(a) Find the set of all points symmetric to the focus F with respect to the tangents of the parabola.

(b) Prove that the set formed by the feet of the perpendiculars dropped from the focus F onto the tangents to the parabola is a straight line parallel to l.

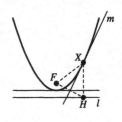

6.12*. (a) Prove that the product of the distances from the foci of an ellipse to any tangent is a constant (i.e., is independent of the particular tangent). ↓

(b) Find the set of points for which an ellipse subtends a right angle (i.e., the set of points where the pairs of tangents to the ellipse meet at right angles).

6.13*. Solve Problem **6.12** (a) for a hyperbola.

6.14*. Solve Problem **6.12** (b) for a parabola.

6.15*. Suppose the trajectory $P_0 P_1 P_2 P_3 \ldots$ of a ray of light inside an elliptic mirror does not pass through the foci A and B (where P_0, P_1, $P_2 \ldots$ are points on the ellipse). Prove that:

(a) If the segment $P_0 P_1$ does not intersect the segment AB, then all the segments $P_1 P_2$, $P_2 P_3$, $P_3 P_4$, $\ldots$, and so on, also do not intersect the segment AB, and these line segments are tangent to a single ellipse with foci at A and B. ↓

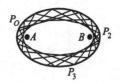

(b) If the segment $P_0 P_1$ intersects AB, then all the segments $P_1 P_2$, $P_2 P_3$, $P_3 P_4 \ldots$, and so on, intersect the segment AB, and the straight lines $P_0 P_1$, $P_1 P_2$, $P_2'P_3$, and so on, are all tangent to a single hyperbola with foci at A and B. ↓

Curves as the envelopes of straight lines

So far, all the curves we have examined—circles, ellipses, hyperbolas, parabolas—arose as sets of points satisfying certain conditions. In the following problems, these curves are generated in a different way: as envelopes of families of straight lines. The word "envelope" simply means that each one of the straight lines of the family is tangent to the curve at some point.

The section of a cone by an arbitrary plane (called a secant plane) not passing through its vertex is an ellipse, a hyperbola or a parabola (Fig. 1). If a sphere touching the secant plane is inscribed in a cone, then the point of tangency will be the focus of the corresponding section, and the directrix will be the line of intersection of the secant plane with the plane of the circle along which the sphere touches the cone.

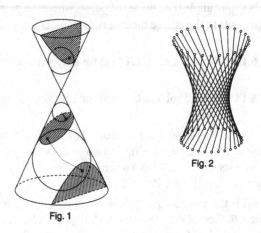

Fig. 1

Fig. 2

The union of all straight lines that are at an equal distance from a given straight line l in space and which make a given acute angle with l is a surface known as a *one-sheet hyperboloid of rotation* (Fig. 2). The same surface can be obtained by rotating a hyperbola around its axis of symmetry l. The tangent plane to the hyperboloid at an arbitrary point intersects the hyperboloid along two straight lines. The remaining plane sections of this surface, as of a cone, are ellipses, hyperbolas and parabolas.

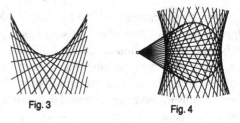

Fig. 3

Fig. 4

76

Fig. 5

Fig. 6

If the points P and N move uniformly along two intersecting straight lines, then the lines PN are either parallel to each other or (in the general case) touch a single parabola (Fig. 3). If the points P and N move uniformly along two skew lines in space, then the union of all the lines PN will be the surface of a hyperbolic paraboloid (saddle-shaped). The tangent plane to the saddle at any point on it intersects it along two straight lines; the remaining plane sections of the saddle are hyperbolas or parabolas. The saddle-shaped surface can also be obtained as the union of all straight lines intersecting two given skew lines l_1 and l_2 and parallel to a given plane (crossing the lines l_1 and l_2).

Figs. 4–6 illustrate problems **6.16** and **6.17**. Note that on our diagrams, only the families of straight lines are drawn; however, the illusion is created that their envelopes—a hyperbola, an ellipse or a parabola, as the case may be—are also drawn on them.

6.16. Suppose we are given a circle with center at O and a point A. Suppose we draw, through each point M on the circle, a straight line perpendicular to the segment MA. Prove that the envelope of this family will be:

(a) A circle, if A coincides with the center O;

(b) An ellipse, if A is located inside the circle;

(c) A hyperbola, if A is located outside the circle. ↓

77

6.17. A straight line l and a point A are given. Through each point M of the given line l, a straight line perpendicular to the segment MA is drawn. Prove that the envelope of this family of straight lines will be a parabola. ↓

These families of straight lines are depicted on pp. 76–77. It is not an accident that all of them form an envelope: indeed, it can be proved that any "sufficiently nice" family of straight lines is either a set of parallel lines, or a set of straight lines passing through a single point, or in the general case, a set of tangents to some curve (the envelope of this family).

Equations of curves

At the beginning of this section we gave geometrical definitions of an ellipse, a hyperbola and a parabola. We can obtain much more information about these curves if we introduce coordinates.

Let us start with the parabola. The analytical definition of a parabola as a graph of the function

$$y = ax^2 \qquad (1)$$

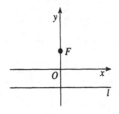

is well known.

We shall show how the geometric definition of a parabola given above results in this equation.

Let the distance from the point F to the straight line l be equal to $2h$. Let us choose a coordinate system Oxy such that the axis Ox is parallel to l and equidistant from F to l, and the axis Oy passes through the point F (the axis Oy will then be the axis of symmetry of the parabola). The equation obtained from the geometric definition of a parabola is easily transformed into (1):

$$\sqrt{x^2 + (y - h)^2} = |y + h|,$$
$$\Updownarrow$$
$$x^2 + y^2 - 2yh + h^2 = y^2 + 2yh + h^2,$$
$$\Updownarrow$$
$$y = x^2/(4h).$$

78

To obtain the form of Equation (1), it suffices to put $a = 1/(4h)$.

The graph of any function of the form $y = ax^2 + bx + c$ is also a parabola. It can be obtained from the parabola $y = ax^2$ by a parallel displacement.

In a similarity transformation $(x, y) \to (ax, ay)$ with coefficient a, the parabola $y = x^2$ becomes the parabola $y = ax^2$. Thus all the parabolas are similar to one another. But parabolas with different values of the parameter a are of course not congruent: the larger the value of a, the "sharper the curvature" of the parabola. Note that one can obtain the parabola $y = ax^2$ from the parabola $y = x^2$ by a contraction (or extension) of one of the coordinate axes, i.e., by the transformation $(x, y) \to (x\sqrt{a}, y)$ or by the transformation $(x, y) \to (x, y/a)$.

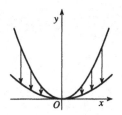

Let us now consider the case of an *ellipse* and a *hyperbola* with foci at A and B. If their axes of symmetry are taken as the axes Ox and Oy of a rectangular coordinate system, then the points A and B will have coordinates $A(-c, 0)$ and $B(c, 0)$, and we will obtain the following equation for an ellipse:

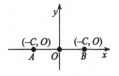

$$\sqrt{(x + c)^2 + y^2} + \sqrt{(x - c)^2 + y^2}$$

$$= 2a \quad \text{(where } a > c\text{).} \qquad (2')$$

By eliminating the radicals, we can express this equation in a more convenient form:

$$\frac{x^2}{a^2} + \frac{y^2}{b^2} = 1, \quad \text{where } b = \sqrt{a^2 - c^2}. \qquad (2)$$

Later on, we will briefly discuss how we can obtain equation (2) from (2').

It can be seen from equation (2) that an ellipse can also be obtained in the following way: take a circle of radius a

$$x^2 + y^2 = a^2$$

and contract it by the ratio a/b towards the axis Ox. Under this contraction, the point (x, y) will be transformed to the point (x, y'), where $y' = yb/a$. Substituting $y = y'a/b$ in the equation of the circle, we get

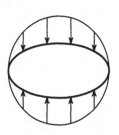

79

the equation of an ellipse: $\frac{x^2}{a^2} + \frac{(y')^2}{b^2} = 1$. Thus you can get an ellipse without using a thread and nails; for example, the shadow cast by a plate, held at some angle on the top of a table, is an ellipse.

Two ellipses are similar to each other if they have the same ratio b/a.

Taking the same coordinate system as in the case of the ellipse, we get the equation of a hyperbola

$$\left| \sqrt{(x+c)^2 + y^2} - \sqrt{(x-c)^2 + y^2} \right|$$

$$= 2a \quad \text{where } a < c, \tag{3'}$$

or after simplification,

$$\frac{x^2}{a^2} - \frac{y^2}{b^2} = 1 \quad \text{where } b = \sqrt{c^2 - a^2}. \tag{3}$$

In order to study the behavior of a hyperbola in the first quadrant $x \geq 0$, $y \geq 0$, let us plot the graph of the function

$$y = \frac{b}{a}\sqrt{x^2 - a^2}.$$

It is clear that this function is defined when $x \geq a$ and increases monotonically. It is not quite so clear—but, nonetheless, true—that as x increases, the hyperbola gets closer and closer to the straight line $y = \frac{b}{a}x$—i.e., that it has this straight line as an *asymptote*.[1]

In fact, the hyperbola has two asymptotes: $y = bx/a$ and $y = -bx/a$.

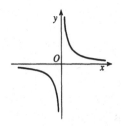

One often encounters another equation whose solution set is referred to as a hyperbola: namely, the equation

$$xy = d \tag{4}$$

(where d is some number $d \neq 0$).

We have to ask ourselves whether this is some other curve or the same curve.

---

[1] More exactly, this means that for any arbitrary sequence x_n tending to infinity, the difference $\left| \frac{b}{a}\sqrt{x_n^2 - a^2} - \frac{b}{a}x_n \right|$ tends to zero. This can be readily proved by using the equality

$$x - \sqrt{x^2 - a^2} = \frac{a^2}{\sqrt{x^2 - a^2} + x}.$$

The curve is of course the same one. To be more precise, the equation $xy = d$ describes a hyperbola with perpendicular asymptotes. The standard equation (3) for such a hyperbola has the form

$$\frac{x^2}{2d} - \frac{y^2}{2d} = 1,$$

but we get equations of different types if we use different coordinate systems. In one case we take the asymptotes of the hyperbola as the coordinate axes; in the other case, we take its axes of symmetry as the coordinate axes. ⟨?⟩.

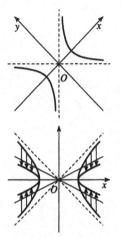

We have shown above how we can obtain an ellipse from the circle $x^2 + y^2 = a^2$ by contraction. In exactly the same way we can obtain the hyperbola $\frac{x^2}{a^2} - \frac{y^2}{b^2} = 1$ (with arbitrary a and b) from the hyperbola with perpendicular asymptotes $x^2 - y^2 = a^2$ by a contraction towards the axis Ox with coefficient a/b.

Two hyperbolas are similar if they have the same ratio b/a, or, equivalently, if they have the same angle 2γ between their asymptotes ($\tan \gamma = b/a$).

A note about the elimination of radicals

Suppose

$$z_1 = \left(\frac{\sqrt{(x+c)^2 + y^2} - \sqrt{(x-c)^2 + y^2}}{2} \right)^2, \quad (3'')$$

$$z_2 = \left(\frac{\sqrt{(x+c)^2 + y^2} + \sqrt{(x-c)^2 + y^2}}{2} \right)^2. \quad (2'')$$

Then, by direct computation, $z_1 + z_2 = x^2 + y^2 + c^2$, $z_1 z_2 = c^2 x^2$; i.e., z_1 and z_2 are the roots of the following quadratic equation:

$$z^2 - (x^2 + y^2 + c^2)z + c^2 x^2 = 0. \qquad (5)$$

The roots of this equation are always nonnegative (since they are both squares), and $z_1 \leq c^2 \leq z_2$ because the quadratic trinomial on the left side of (5) is nonnegative when $z = 0$ and nonpositive when $z = c^2$.

Note that if $z \neq 0$, $z \neq c^2$, equation (5) may be rewritten as follows:

$$\frac{x^2}{z} + \frac{y^2}{z - c^2} = 1.$$

Let $a^2 < c^2$, $a > 0$ and (3′) hold. Then $z = a^2$ is the smaller root of (5), $0 < z < c^2$, and therefore the equation

$$\frac{x^2}{a^2} + \frac{y^2}{a^2 - c^2} = 1 \qquad (6)$$

(provided that $0 < a < c$) is equivalent to (3′). Setting $b = \sqrt{c^2 - a^2}$, we see that (3) $\Longleftrightarrow$ (3′).

Suppose $a^2 > c^2$, $a > 0$, and (2′) hold. Then $z = a^2$ is the larger root of (5), $z > c^2$. Hence equation (6) is equivalent to (2′) provided that $a > c$. Setting $b = \sqrt{a^2 - c^2}$, we obtain (2) $\Longleftrightarrow$ (2′).

This proof illustrates a method frequently used for eliminating radicals: consider, together with a given expression, its conjugate expression, which differs from the original only in the sign before the radical.

The end of our "alphabet"

Finally, let us consider one more function on the plane whose map of level curves includes all three types of curves appearing in this section. This will give us the last proposition of our "alphabet."

N. *Suppose we are given a point F and a straight line l not containing the point F. Given any point p, consider the ratio of its distances from F and l. The set of points for which that ratio is equal to a constant k is an ellipse (when $k < 1$), a parabola (when $k = 1$) or a hyperbola (when $k > 1$).*

Let us prove this. Let us introduce a coordinate system as we did above in the section on the parabola. The equation of the required set is

$$\frac{\sqrt{x^2 + (y - h)^2}}{|y + h|} = k.$$

When $k = 1$, as we have already seen, this is equivalent to the equation of the parabola $y = ax^2$, where $a =$

$1/(4h)$. When $0 < k < 1$, it can be reduced to the form

$$\frac{x^2}{a^2} + \frac{(y-d)^2}{b^2} = 1 \quad \text{(an ellipse)}, \qquad (7)$$

and when $k > 1$, to the form

$$\frac{(y-d)^2}{b^2} - \frac{(x)^2}{a^2} = 1 \quad \text{(a hyperbola)}, \qquad (8)$$

where in both cases

$$a = 2kh/\sqrt{|k^2-1|}, \quad b = 2kh/|k^2-1|,$$

and

$$d = h(k^2+1)/(k^2-1).$$

Equations (7) and (8) are obtained from the standard equations (2), (3) by a parallel displacement and an interchange of x and y. Now, the foci of the curves lie on the axis Oy, and the centers are displaced to the point $(0, d)$. It may be verified that the point F is the focus not only of the parabola but also of all the ellipses and hyperbolas. The straight line l is called their directrix.

Thus, we have seen that the set of level curves of the function

$$f(M) = \rho(M, F)/\rho(M, l)$$

consists of ellipses, hyperbolas, and a single parabola.

We might have guessed that these curves would be "conic sections" (see pp. 67 and 76—77) by reasoning as follows. Consider two functions on a plane: $f_1(M) = \rho(M, F)$ and $f_2(M) = k\rho(M, l)$. The graph of the first function is the surface of a cone; the graph of the second consists of two inclined half planes (k is the tangent of the angle of inclination of these half planes to the horizontal). The intersection of these two graphs is an ellipse, a parabola or a hyperbola. The projections onto the horizontal plane of these curves on an inclined plane give the required sets:

$$\{M: f_1(M) = f_2(M)\} = \{M: \rho(M, F)$$
$$= k\rho(M, l)\}.$$

When projected, the form of the curve changes, as if contracted towards the straight line l (in the ratio

$\sqrt{k^2 + 1}$). Hence, our required curves are also ellipses, hyperbolas and a parabola.

As we have already repeatedly found, the curves discussed in this chapter—the ellipse, the hyperbola and the parabola—possess many common or very similar properties. The relationship between these curves has a simple algebraic explanation: all of them are given by quadratic equations. Of course, the standard equations of these curves (1), (2), (3), (4), i.e.,

$$y = ax^2, \quad \frac{x^2}{a^2} + \frac{y^2}{b^2} = 1,$$

$$\frac{x^2}{a^2} - \frac{y^2}{b^2} = 1, \quad xy = d,$$

are obtained only in specially selected coordinate systems. If the coordinate system is chosen in some other way, the equations may be more complicated. However, it is not difficult to prove that in any arbitrary coordinate system, the equations of these curves have the form

$$ax^2 + bxy + cy^2 + dx + ey + f = 0 \qquad (9)$$

(where a, b, c, d, e, f are certain numbers and $a^2 + b^2 + c^2 \neq 0$).

Remarkably, the converse is also true: any equation of the second degree $p(x, y) = 0$, i.e., any equation of the form (9) determines one of these curves. Let us formulate the theorem more precisely.

Equation (9) defines an ellipse, a hyperbola or a parabola only if the left-hand side does not decompose into factors (if it did, we would get a pair of straight lines) *and assumes values of both signs* (if not, we would get a single point, a straight line or the empty set). The origin of the general name "quadratic curves" for ellipses, hyperbolas and parabolas becomes clear from this.

This important algebraic theorem for the second degree equations is very helpful when looking for point-sets satisfying a geometric condition: if we find that in some coordinate system this condition is expressed by a second-degree equation, then the required set is an ellipse, a hyperbola, or a parabola. (Of course, in the

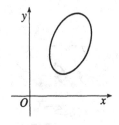

case of degeneracy, we may get a pair of straight lines, a circle which is a particular case of an ellipse, a single point, etc.). One merely has to determine their dimensions and position in the plane (the foci, the center, the asymptotes, etc.).

6.18. Let l_1 and l_2 be two mutually perpendicular lines, and let p be their point of intersection. Find the set of points q such that the sum of the distances from q to l_1 and l_2 is c units greater than the distance from q to p.

6.19. Given a straight line l and a point A in a plane, find the set of points:

(a) the sum of whose distances from A and l is equal to c;

(b) the difference of whose distances from A and l is equal (in absolute value) to c;

(c) the ratio of whose distances from A and l is less than c, where c is a positive constant.

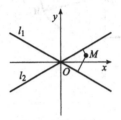

6.20. Let l_1 and l_2 be two intersecting straight lines. Let d be a constant. Find the set of points M such that

(a) $d = \rho^2(M, l_1) + \rho^2(M, l_2)$,

(b) $d = \rho^2(M, l_1) - \rho^2(M, l_2)$.

Draw the map of the level curves of the corresponding functions:

(a) $f(M) = \rho^2(M, l_1) + \rho^2(M, l_2)$,

(b) $f(M) = \rho^2(M, l_1) - \rho^2(M, l_2)$.

6.21. Given a point F and a straight line l in a plane, draw the map of the level curves of the functions:

(a) $f(M) = \rho^2(M, F) + \rho^2(M, l)$,

(b) $f(M) = \rho^2(M, F) - \rho^2(M, l)$.

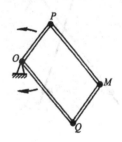

6.22. The vertex O of a hinged parallelogram $OPMQ$ is fixed while the sides OP and OQ rotate with angular velocities which are equal in magnitude and opposite in direction. Along what line does the vertex M move?

□ Let $|OP| = p$, $|OQ| = q$. Since OP and OQ rotate in opposite directions, they will coincide at some point in time. Take this point as the initial point in time $t = 0$, and the coincident lines as the axis Ox (We take the origin of the coordinate system to be the point O).

85

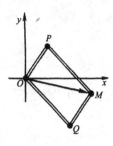

Let the sides OP and OQ rotate with angular velocity ω. Then the coordinates of the points P, Q will, at time t, be equal to

$$(p \cos \omega t, \; p \sin \omega t),$$
$$(q \cos \omega t, \; -q \sin \omega t), \quad \text{respectively.}$$

Hence, the coordinates of the point $M(x, y)$ will be

$$x = (p + q) \cos \omega t,$$
$$y = (p - q) \sin \omega t$$

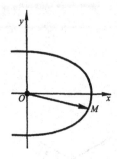

(since $\overrightarrow{OM} = \overrightarrow{OP} + \overrightarrow{OQ}$). Therefore, the point M describes an ellipse

$$\frac{x^2}{(p+q)^2} + \frac{y^2}{(p-q)^2} = 1. \quad \square$$

In the solution to this problem, we obtained the ellipse as a set of points (x, y) of the form

$$x = a \cos \omega t, \quad y = b \sin \omega t \qquad (10)$$

(where t is an arbitrary real number). Equations of this type, which express the coordinates (x, y) in terms of an auxiliary parameter t, are called *parametric* equations. In this particular case, the variable parameter t represents time.

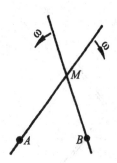

6.23*. In a plane, two straight lines passing through two fixed points A and B rotate about these points with equal angular velocities. What line does their point of intersection M describe if the lines rotate in opposite directions? ↓

6.24*. Find the set of points M in a plane for which $\widehat{MBA} = 2\widehat{MAB}$, where AB is a given segment in the plane. ↓

6.25*. (a) Consider all the segments that cut off a triangle of area S from a given angle. Prove that the midpoints of these segments lie on a hyperbola H whose asymptotes are the sides of the angle. ↓

(b) Prove that all these segments touch the hyperbola H. ↓

86

(c) Prove that the segment of a tangent to the hyperbola cut off by the asymptotes is bisected at the point of tangency. ↓

6.26*. (a) Suppose we are given an isosceles triangle ABC ($|AC| = |BC|$).

Find the set of points M in a plane such that the distance from M to the straight line AB is equal to the geometric mean of the distances from M to the lines AC and BC.

(b) Three straight lines intersecting each other form an equilateral triangle. Find the set of points M such that the distance from M to one of these straight lines is equal to the geometric mean of the distances from M to the other two.

6.27*. A rectangle $ABCD$ is given in a plane. Find the set of points M such that $\widehat{AMB} = \widehat{CMD}$.

Algebraic curves

Obviously, the sets of points which one may meet in geometrical problems are not limited to straight lines and quadratic curves. Let us give two examples.

The set of points, the product of whose distances from two given points F_1 and F_2 is equal to a given positive number p, is called an *oval of Cassini*. A whole family of these curves—the family of level curves of the function

$$f(M) = \rho(M, F_1)\rho(M, F_2)$$

is shown in the figure.

Equations of these curves may be written as follows:

$$((x - c)^2 + y^2)((x + c)^2 + y^2) = p^2.$$

An oval of Cassini has the particularly interesting form of a "figure eight," when $p = c^2$. When $p < c^2$, the curve consists of two separate parts surrounding the points F_1 and F_2.

Here is the other example. Let a point F and a straight line l be given. Denote the distance of a point M from the point of intersection of the straight lines

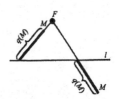

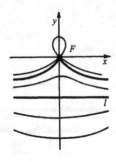

FM and l by $q(M)$. The set of points $\{M : q(M) = d\}$ is called the *conchoid of Nicomedes*. Its equation in the coordinate system where F is the origin and l is given by the equation $y + a = 0$ is expressed as follows:

$$(x^2 + y^2)(y + a)^2 - d^2 y^2 = 0.$$

In general, the curve given by the equation $P(x, y) = 0$, where $P(x, y)$ is a polynomial in x and y, is called an *algebraic curve*. The degree of the polynomial P (provided that it does not factor) is called the *order* of the curve. Thus, the oval of Cassini and the conchoid are curves of the fourth order. It is already clear from these two examples that algebraic curves (of order higher than 2) may look somewhat peculiar: they may possess singular points (cusps, as the conchoid has when $a = d$, or points of self-intersection) and the form of these curves may change significantly when the parameters are changed. We shall meet some new curves in the next chapter.

CHAPTER 7

Rotations
and
Trajectories

In this chapter we present some remarkable curves that are naturally generated as trajectories of points on a circle rolling along a straight line or along another circle. The most interesting properties of these curves are connected with tangents. We will start by investigating cycloids, which are the paths traced by a single point on a circle as the circle rotates along another curve. The reader may recall that, at the end of the Introduction, we revisited Problem **0.1** and encountered a curve realized as the envelope of a family of lines. This envelope was a curve with four cusps, called an astroid. We will examine this fact in greater detail here, and we will also see why a spot of light in a cup formed by reflected rays has a characteristic singularity, a cusp. The devotee of classical geometry will find out about the connections between the nine-point circle of a triangle, its Wallace–Simson lines and their envelope, the Steiner deltoid, which is a cycloid with three cusps.

We shall first study one of the simplest cycloids.

The cardioid

Usually, this curve is defined as the path of a point moving in the following way: Given a stationary circle, suppose another circle of the same radius rolls without slipping around the stationary circle. Fix a point on the moving circle. The path traced by this particular point on the moving circle is called a *cardioid*.

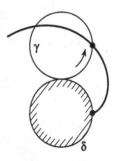

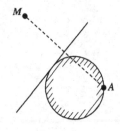

It is possible to give other geometrical definitions of a cardioid. We shall give two of them in the form of an exercise for the reader.

7.1. Prove the following statements:

(a) Let A be a particular point on a given circle. Consider the set of points symmetric to A relative to all the possible tangents to this circle. Prove that this set of points is a cardioid;

(b) As before, let A be a particular point on a given circle. Consider the set consisting of the feet of the perpendiculars dropped from A onto all possible tangents to the circle. Prove that this set of points is a cardioid.

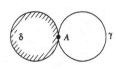

□ (a) Consider a circle γ which touches a given circle δ at the point A and has the same radius as δ. Suppose the circle γ rolls around the circle δ; let M denote the point on the moving circle which, at the initial moment in time, coincides with the point A. Let us follow the path of the point M.

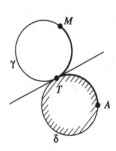

We assume that the circle γ rolls without slipping. This means that if T is the variable point of contact between the circles, then, at every instant, the lengths of arcs AT and MT are equal. Hence the point M is symmetric to the point A with respect to the tangent drawn through the point T.

In a single revolution, the point T runs around the whole of the circumference of the circle δ, and M around the entire cardioid.

(b) Clearly, we can obtain this set from the one mentioned in (a) by a similarity transformation with coefficient 1/2 and center A. Hence it is also a cardioid of half the size of the cardioid in (a). □

Using Problem **7.1**, we can plot as many points of the cardioid as we please, and thereby draw it quite accurately. It is a closed curve which has a characteristic singularity—a cusp—at the point A. The shape of this curve resembles the cross-section of an apple, somewhat in the shape of a heart, from which its name comes (*Kardia* means "heart" in Greek).

The next beautiful definition of a cardioid, in which it is generated as an "envelope of circles," also follows from Problem **7.1**.

90

7.2*. Suppose we are given a circle γ and a point A lying on it. Prove that the union of all the circles that pass through the point A, and whose centers lie on the circle γ, is a region bounded by a cardioid. ↓

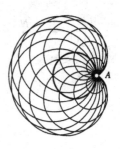

Addition of rotations

We are now going to discuss ways of determining the geometric properties of curves with the help of kinematics. The cardioid will serve as an example. But before proceeding further, let us discuss the last sentence of the solution to Problem **7.1** (a).

We said that the point T returns to the initial point A *after one revolution*. As we are dealing with several different rotations, this phrase needs to be made more precise: what is a "revolution," i.e., exactly what rotation are we talking about?

What we mean is that the center P of the moving circle γ (and therefore the point of tangency T) makes one revolution. But the circumference of the circle γ itself (we can visualize it better as a circular plate) rotates about its center P quite quickly. Let us study this motion in greater detail.

7.3. Suppose the center P of the moving circle γ, rolling along a stationary circle δ of the same radius, makes one revolution around δ. How many revolutions will the circle γ make about its center P during this time?

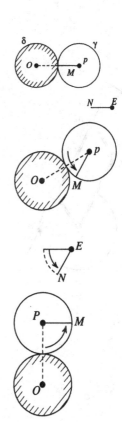

□ In order to follow the rotation of the circle γ, let us draw a radius PM in the circle γ. Let E be a fixed point in the plane, and let EN be a segment such that $\overrightarrow{EN} = \overrightarrow{PM}$. Our question is: How many revolutions will the segment EN make about its endpoint E while the segment OP rotates through $360°$? In other words, what is the ratio of the angular velocities of these segments?

To answer this question, it is sufficient to consider two different positions of the moving circle. One can see from the figure that when the radius OP turns through $90°$, the segment EN turns through $180°$. Continuing further in the same way, we see that when the radius OP turns through $360°$, the segment EN will turn through 720°; i.e., it will make two complete revolutions (the ratio of angular velocities is equal to 2). This gives us the answer to Problem **7.3**. ☐

If we take the center O of the stationary circle as the point E in the solution of Problem **7.3** and mark off from it the segment $\overrightarrow{OQ} = \overrightarrow{PM}$, then we obtain the parallelogram $OPMQ$.

In the uniform rolling of the circle γ around δ, the vertex O is motionless and the sides OP and OQ rotate with the angular velocities ω and 2ω, respectively (in the same direction). Thus, we obtain another definition of the cardioid using the convenient model of a hinged parallelogram:

If the sides OP and OQ (where $|OP| = 2|OQ|$) rotate about the point O with angular velocities ω and 2ω, the locus of the fourth vertex M of the parallelogram $OPMQ$ is a cardioid.

It is now easy to give one more method for the construction of a cardioid. We will also deduce a few more of its fascinating properties.

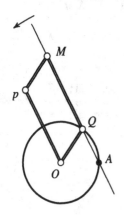

7.4. Suppose we are given a circle δ of radius r and a point A lying on it. If, on every straight line l passing through the point A, we mark off from the point of intersection Q of l and δ ($A \neq Q$) the segment QM of length $2r$, then the set of all points M thus obtained will be a cardioid.

☐ For every position of the straight line l, we may construct a parallelogram $OPMQ$, where Q and M are as stated in the problem. Then, if the straight line l rotates about the point A with an angular velocity ω, the sides OP and OQ of the parallelogram will rotate with exactly the necessary velocities ω and 2ω (according to the theorem about a ring on a circle in Chapter 1), and so the point M will describe a cardioid. ☐

Try to construct a cardioid on a large sheet of paper using problems **7.1** and **7.4** and convince yourself that you obtain the same curve. Perhaps the second method is even more convenient. Note that in Problem **7.4** we may mark off the segment QM of length $2r$ from the point Q in either direction. From this, we obtain two points M_1 and M_2 of the cardioid. They correspond to two opposite positions of the hinged parallelogram (if the point Q makes one full revolution and returns to the initial point, then the side OM will turn through $180°$ and M_1 will coincide with the point M_2). This circumstance leads to the following property.

7.5. Suppose we are given a cardioid with its cusp at the point A. Prove that any chord M_1M_2 of the cardioid passing through A has length $4r$, and that the midpoint of the chord lies on the stationary circle (of radius r) that generates the cardioid.

Here are two more problems which use the second method of constructing a cardioid.

7.6. A stick of length $2r$ moves in a vertical plane so that its lower end rests against the bottom of a hole in the ground whose vertical cross-section is a semicircle of radius r. The stick rests against the edge of the hole. Prove that the free upper end of the stick moves along a portion of a cardioid.

7.7. A hoop of radius $2r$ rolls, without slipping, around the outside of a stationary circle of radius r. Prove that the locus of a fixed point on the hoop is cardioid.

□ One solution to this problem may be obtained if we compare the problem with Copernicus' Theorem **0.3**. Here, in fact, we are dealing with the same two circles, but the internal circle of radius r is fixed, and the external circle of radius $2r$ rolls around it. In this situation, Copernicus' Theorem shows that if we fix a stick to the hoop along the diameter M_1M_2, then, while rolling, the stick passes through a fixed point A of the stationary circle. At the same time, the midpoint Q of the stick M_1M_2 moves around the stationary circle δ, and $|M_1Q| = |QM_2| = 2r$. Hence we arrive at Problem **7.4**, and we can see that the points M_1 and M_2 move on the same cardioid.

One can reason in a somewhat different way, making the problem analogous to that of the hinged parallelogram. Let M be the point of the hoop we are following and Q its (variable) center. We shall construct the parallelogram $OPMQ$. If the link OQ of the parallelogram rotates with angular velocity 2ω, then the hoop, and with it the link QM, rotate with angular velocity ω. □

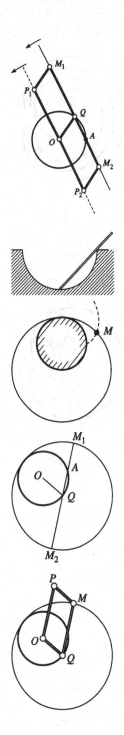

93

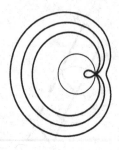

The curve we have just been considering, the cardioid, is included in a natural way in the family of curves called *conchoids of a circle* or the *limaçon of Pascal*. Consider the statement of Problem **7.4**: suppose that on the straight line l passing through the point A, we mark off a segment QM of some constant length h (in either direction). Then we get one of these curves for every $h > 0$. For $h = 2r$, the curve will be a cardioid. It turns out that we can give a kinematic definition of the limaçon of Pascal for every h. We do this in the next problem.

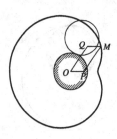

7.8. (a) Prove that the vertex M of a hinged parallelogram, whose vertex O is fixed and whose sides OP and OQ rotate with angular velocities 2ω and ω, respectively, describes a limaçon of Pascal.

(b) A circle of radius r is fixed in a plane. Around it rolls a circle of radius r with a moving plane rigidly fixed to it. Prove that every point of this plane describes a limaçon of Pascal.

(c) Repeat part (b), but suppose that instead of a moving circle of radius r, we have a loop of radius $2r$ encircling the stationary circle.

Now let us investigate some problems which require us to look at the addition of rotations where there is a different ratio between the velocities than we had in the case of the cardioid. We will be reminded of some of the other cycloids shown in the figures on pp. 95–96.

7.9. A circle of radius (a) $R/2$, (b) $R/3$, (c) $2R/3$ is rolling around the outside of a stationary circle of radius R. In each case, how many revolutions will the circle make while its center describes one revolution about the center of the stationary circle? ↓

7.10. Solve the same problem, but with the circle rolling around the inside.

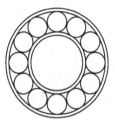

7.11. Ball bearings with 2mm diameters are located between the axle of a bearing 6 mm in diameter and its stationary ball race of 10 mm in diameter. When the axis rotates, the ball bearings roll around the axle and the balls race without slipping. Find out with what angular velocity (a) the ball bearings rotate, and (b) their centers run about the center of the bearing if the axle rotates with an angular velocity of 100 revolutions per second.

7.12. Gears that propel a grindstone are arranged as is shown in the diagram. Find the ratio of the radii of the moving wheels for which the smaller wheel (the grindstone) will revolve 12 times faster than the handle OQ which sets it in motion.

Consider two points on a circle as the circle rolls around another circle. It is clear that they must describe congruent paths. In particular, it is possible for these two paths to coincide: the two points can move along the same curve, one following the other. This was the case, for instance, in the solution of Problem **7.7**, where we saw that diametrically opposite points of a hoop described the same cardioid. We could have convinced ourselves of this by simply noticing that the paths of these points have their cusps at the same point of the stationary circle. We can use similar observations in the following problems.

Fig. 1

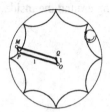

Fig. 2

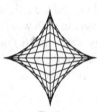

Fig. 3

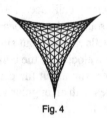

Fig. 4

A *k-cycloid* is the curve described by the vertex M of a hinged parallelogram $OPMQ$, whose vertex O is fixed and whose links OP and OQ rotate about O, where the ratio ω_{OP}/ω_{OQ} of the angular velocities is equal to k and the ratio $|OP|/|OQ|$ of the lengths of the links is equal to $1/|k|$ $(k \neq 0, +1, -1)$.

If two points L and N move uniformly around a circle, so that the ratio ω_L/ω_N of their angular velocities is equal to k, then the envelope of the straight lines LN will be a k-cycloid (**7.19**).

The shapes of a k-cycloid and a $(1/k)$-cycloid coincide (**7.14**).

The k-cycloid may also be defined as the locus of a point of a circle of radius r which rolls around another circle of radius $|k-1| \cdot r$ without slipping (externally when $k > 1$ and internally when $k < 1$).

96

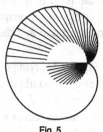

Fig. 5

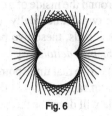

Fig. 6

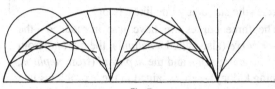

Fig. 7

Usually k-cycloids are called *epicycloids* when $k > 0$, and *hypocycloids* when $k < 0$. In diagrams 1–6, k-cycloids are depicted for $k = 3/8$, $-1/7$, -3, -2, $1/2$, and 3. The last four have special names: the *astroid*, *Steiner deltoid*, *cardioid*, and *nephroid*. Several families of segments related to these curves are shown in diagrams 3–6. All the segments in each diagram have equal lengths (**7.4**; the theorem on two circles on p. 99; **7.21**).

In the last diagram 7, the locus of a point of a circle rolling along a straight line is shown. This curve is known as the *cycloid*. The envelope of the diameters of the rolling circle is a cycloid of half the size (the theorem on two circles).

97

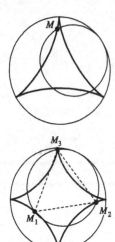

7.13. (a) Suppose that γ, a circle of radius $2R/3$, rolls around the inside of a circle of radius R. If M_1 and M_2 are two diametrically opposite points on γ, prove that as γ rolls, these two points trace out one and the same *Steiner deltoid*. ↓

(b) Prove that three points M_1, M_2 and M_3 lying on a circle of radius $3R/4$ at the vertices of an equilateral triangle will describe the same curve, an *astroid*, if the circle is rolled around the inside of a circle of radius R.

(c) Solve the same problem as in (b), but suppose the radius is $3R/2$ instead of $3R/4$. In this case, instead of an astroid, we get a *nephroid* (and the movable circle encircles the stationary one like a hoop).

The three curves we have just introduced—the *Steiner curve* (also called a *deltoid*), the *astroid* (from *astra*, meaning star), and the *nephroid* (from *nephros*, meaning kidney)—are obtained in these problems in a somewhat different way from the way they are defined on pp. 96–97.

We have already seen, from the example of the cardioid, that a curve may be obtained as the paths of points on two different circles rolling around the one stationary circle (Compare the first definition of the cardioid and Problem **7.7**. In the first case, the center of the moving circle is the vertex P of a hinged parallelogram $OPQM$, and in the second case, the vertex Q). The following problem shows us what ratios between the radii of the circles we must take to obtain congruent paths.

7.14*. (a) Prove that a point on a circle of radius r, rolling around the outside of a stationary circle of radius R, and a point on a circle (hoop) of radius $R+r$, surrounding the circle, describe congruent paths.

(b) Prove that a point on a circle of radius r, rolling around the inside of a circle of radius R, and a point on a circle of radius $R - r$, rolling inside the same fixed circle, describe congruent paths. ↓

To solve these problems, we have to learn how to calculate the ratios of the velocities of quite complicated rotations. We shall discuss how to do this below, but now let us go on to the most interesting properties of cycloids: the properties of their tangents.

A theorem on two circles

We will formulate a curious rule which allows us to describe the family of tangents to the trajectory of the point M on a circle of radius r which rolls without slipping along a curve γ. Let us roll a circle of radius $2r$ along the same curve γ, and suppose that a diameter KL of this circle (considered fixed relative to the circle), as shown, is positioned in such a way that at some instant its endpoint K and the point M coincide at the point A on the curve γ. It so happens that in this case, at any point in time, the *diameter KL is tangent to the path of the point M*. In other words, the *path is the envelope of all the positions of the diameter KL.*

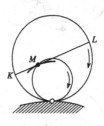

We have called this very convenient rule the *"theorem on two circles."* We shall discuss its proof later on, but first let us make things a little clearer. If we roll the two circles mentioned in the theorem simultaneously, so that their points of tangency with the curve γ always coincide, then the smaller circle will roll around the bigger one without slipping. Then, from Copernicus' Theorem, the point M will move along a fixed diameter KL of the bigger circle. Our theorem on two circles asserts that the straight line KL will be tangent at the point M to the locus of this point M.

Let us move on to the examples. Let us begin with the family of curves which we spoke about in the introduction to the book. Assume that a circle of radius r with the point M marked rolls around the inside of a circle of radius $R = 4r$. Together with it, let us roll a circle of radius $2r$ along with its diameter KL. (At the initial moment, the points K and M coincide with the point A on the stationary circle) According to Copernicus' Theorem, the endpoints of the diameter KL slide along two mutually perpendicular diameters AA' and BB' of the stationary circle. At the same time, according to the theorem on two circles, the diameter KL, as it moves, is tangent to the trajectory of the point M, i.e., the *envelope of the straight lines KL is an astroid with cusps at the points A, B, A', B'.*

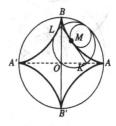

The next problem is about the cardioid.

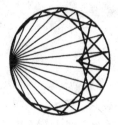

7.15*. A point B is given on a circle. From B, a ray of light falls on any arbitrary point on the circle and is then reflected from the circle (the angle of incidence with the circle's tangent is equal to the angle of reflection). Prove that the envelope of the reflected rays is a cardioid.

□ Let us denote the center of the "reflecting" circle by O and the point diametrically opposite the point B by C. Suppose the ray BP, after being reflected at the point P, arrives at the point N of the segment BC (we consider for the time being that $\widehat{PBC} \le 45°$). Then $\widehat{PNC} = \widehat{BPN} + \widehat{PBN} = 3\widehat{PBC}$. This means that if we rotate the ray BP with an angular velocity ω, then the reflected ray will rotate with an angular velocity 3ω, and the point of reflection P will move around the reflecting circle with an angular velocity 2ω (according to the "theorem about the tiny ring" from Chapter 1). Clearly, this will also be the ratio when $\widehat{PBC} > 45°$.

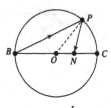

We can get the family of straight lines PN in the following way. Let us roll a circle of radius $2r$, together with its diameter KL (which, at the initial moment, lies along the straight line BC) around a fixed circle of radius $|OB|/3$ with its center at O. If the center P of the moving circle rotates with angular velocity 2ω around the circle of radius $3r$ with center O, then the diameter KL will rotate with an angular velocity 3ω ⟨?⟩ — just as the reflected ray did.

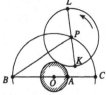

By the theorem about two circles, the envelope of the family of straight lines KL will be the trajectory of the point M of the circle of radius r rolling around a circle of the same radius r with center O; i.e., a cardioid. At the initial moment, the point M coincides with the point A, dividing the segment BC in the ratio 2:1. This point will be the cusp of the cardioid. □

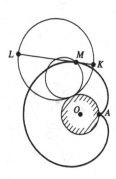

We often see this "cusp" in the form of a spot of light formed by reflected rays at the bottom of a cup or a saucepan inclined to incident rays from a lamp or the sun. However, in such cases it is more natural to consider the pencil of incident rays as being parallel and not coming from a single point on the circle. We do not then get a cardioid, but another known curve, with a similar cusp.

7.16*. Prove that if a collection of parallel rays falls on a semicircular mirror (as shown in the diagram), then the reflected rays are tangent to half a nephroid.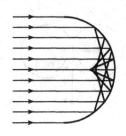

If the mirror were parabolic, then, as we know from Chapter 6, the reflected rays would come together at a single point, the focus of the parabola. This comparison gives rise to the other name for the nephroid: the *focal line of a circle*.

7.17. Find the set of points determined by the envelope of diameters of a circle of radius r, as the circle rolls:

(a) around the outside of a circle of radius r;

(b) around the inside of a circle of radius $3r/2$.

A few more interesting problems about families of tangents appear below, but first we will discuss the kinematic concepts used in the solution of the last few problems and in the proof of the theorem on two circles.

Velocities and tangents

There are more convenient ways to determine the ratios of the angular velocities in these complicated rotations than the quite primitive method we used in solving Problem **7.4**. First of all, there is the rule for adding angular velocities, which is similar to the rule for the addition of linear velocities when changing to a new frame of reference.

Let us take angles (and angular velocities) corresponding to counterclockwise rotations as being positive, and angles and rotations in a clockwise direction as being negative.

Then if the straight line l_2 is turned relative to the straight line l_1 through angle φ' and l_3 is turned relative to l_2 through an angle φ, then l_3 turns with respect to l_1 through the angle $\varphi + \varphi'$.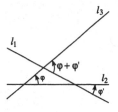

Thus, if *the figure γ_2 rotates with respect to the "fixed" figure γ_1 with an angular velocity ω', and γ_3 with respect to γ_2 with an angular velocity ω, then γ_3 rotates with respect to γ_1 with angular velocity $\omega + \omega'$.* We mostly deal with rotations of circles, so we shall assume that some radius is marked on each of them in order to follow their rotations more easily.

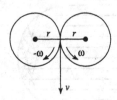

Let us show how we can apply this rule. First, consider two circles of radius r whose centers are fixed at a distance $2r$ from each other. If the circles rotate without slipping, then their angular velocities are equal in value and opposite in sign: the first has angular velocity $-\omega$ and the second has angular velocity ω. This is because the linear velocities of the points of tangency of the two circles are equal (the fact that the circles rotate without slipping is used here). Recall that if a point M is located at a distance r from the center of a circle that rotates with angular velocity ω, then the linear velocity of M is given by $v = \omega r$. From the equality of the *linear* velocities we get the equality of the *angular* velocities of the circles (in absolute value).

Now let us pass to a reference frame fixed to the first circle. We then have to add ω to all the angular velocities: the angular velocity of the first circle will be 0 while the angular velocity of the second circle will be 2ω. We have already seen this in Problem **7.4**.

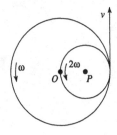

Consider another example. Suppose that the r is the distance between the centers O and P of the mutually tangent circles of radii $R = 2r$ and r, respectively (for the time being, let us take their centers as fixed). Their angular velocities will be ω and 2ω, respectively (the ratio of these values is inversely proportional to the ratio of the radii). In a reference frame fixed to the smaller circle, the angular velocity of the larger circle is $(-\omega)$ and the angular velocity of the smaller circle is 0 (this was the motion which we spoke about in Copernicus' Theorem **0.3**). In a reference frame fixed to the larger circle, the angular velocities of the larger and smaller circle are 0 and ω, respectively (see Problem **7.7**).

When determining angular velocities, it is possible, however, to avoid the introduction of a rotating reference frame. To do so, we must clarify how to find the (linear) velocities of the points on a rolling circle (a wheel). This question is of great importance in the next section, which deals with tangents to cycloids.

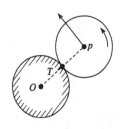

Thus, we return to the first example: let us consider some position of a circle of radius r, rolling around a circle of the same radius; denote by T the point on the moving circle coinciding at the moment considered with the point of tangency of the circles. Its velocity is equal to 0 (since the rotation is without slipping). How do we find the velocities of the other points?

For this, let us apply the following *theorem of Mozzi*:

At any point in time, the *velocities of the points of a solid plate which moves in a plane are either those of a body in translation* (i.e., are all equal in value and have the same direction) *or those of a rotating body*, i.e., the linear velocity of some point T is equal to zero and the linear velocity of every other point M is equal in magnitude to $|MT|\omega$ (where ω is the angular velocity of the plate) and is perpendicular to the segment MT. This last case, in particular, applies to a rolling circle, and the point of tangency plays the role of the point T ("the instantaneous center of rotation"). (This will be true even for an irregular wheel rolling on a bumpy road.) Making use of this, we can find the ratio between the angular velocity ω_1 of the rolling wheel and the angular velocity ω_2 with which its center P rotates about the center O of the stationary circle. To do this, we express the linear velocity of the point P in two different ways: on the one hand, its value is equal to $2r\omega_2$. On the other hand, it is equal to $r\omega_1$, since T is the instantaneous center of rotation. Hence, $2r\omega_2 = r\omega_1$, and so $\omega_1 = 2\omega_2$.

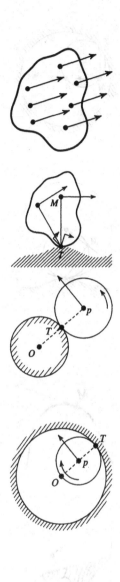

Suppose that a circle of radius r rolls around the inside of a circle of radius $2r$ in such a way that its center moves (around a circle of radius r) with the angular velocity $\omega_2 > 0$. The same reasoning as above allows us to deduce the following: Denote the angular velocity of the circle by ω_1 and note that $\omega_1 < 0$. Expressing the velocity of the point P in two different ways, we get $|\omega_1 r| = |\omega_2 r|$, giving $\omega_1 = -\omega_2$.

Similar reasoning helps us when studying other complex rotations.

But what is particularly important is that Mozzi's theorem allows us to find the *direction* of the velocity at every point of the figure: the velocity of the point M is directed orthogonally to the segment MT joining M with the instantaneous center of rotation T.

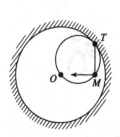

We shall now give one more proof of Copernicus' Theorem. Let M be a point on a circle of radius r which rolls inside a circle of radius $2r$ with center O.

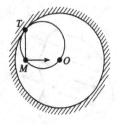

At any point of time, the velocity of the point M is perpendicular to the segment TM, where T is the point of contact of the circles (and the instantaneous center of rotation of the smaller circle). Thus, the velocity of the point is always directed along the straight line MO (since T and O are diametrically opposite points on the smaller circle). Therefore, the point M moves along a diameter of the larger circle, which is just what Copernicus' Theorem asserts.

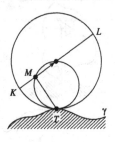

We now give a proof of the *theorem on two circles*. Let us simultaneously roll two circles of radii r and $2r$ along the curved (or straight) line γ. Let M and K be points on them which coincide at the initial moment with the point A of γ, and let T be the common instantaneous center of rotation of the two circles (their point of contact with γ). The velocity of the point M is directed perpendicularly to the segment MT.

Hence, the *velocity of the point M is directed along the diameter of the larger circle*: that is, M lies on a certain diameter KL of this circle, and in its motion, the straight line KL touches the path of the point M. This is just the theorem on two circles.

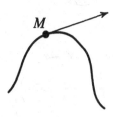

Note that here we have looked at the definition of the tangent to a curve in a new way. The *tangent at the point M to the path of a moving point is the straight line passing through the point M on the path whose direction coincides with the direction of the velocity at the given point M*.

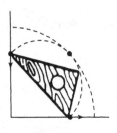

We shall not give a proof of Mozzi's theorem, but we shall point out its geometrical analogue: any displacement of a plane which can be realized without turning the plane over onto the other side (that is, by a direct isometry), is either parallel displacement or rotation about some point T (Chasles' theorem). In connection with Mozzi's theorem, we stress one more idea. In the case of the most general movement of a plate in a plane, the instantaneous center T changes its position not only in the stationary plane, but also in the moving one (the plate) during the process of movement. In each case it describes some curve; one is called the *fixed centrode* and the other the *moving centrode*. For instance, during the rolling of a wheel along a road, the fixed centrode would be the road and

104

the moving centrode would be the rim of the wheel. A well-known kinematics theorem states that for every "smooth" enough motion of a plane, i.e., motion without "jerks," *the moving centrode rolls along the fixed one without slipping*, and at each moment, their point of contact is the instantaneous center of rotation.

Thus the general motion of a plate in a plane reduces to the rolling of an irregular wheel on a bumpy road. From this point of view, the subject matter of our section could be summarized as the study of motions for which both centrodes are circles. With that, we come to the end of our digression into kinematics. We are now equipped to set about discovering some of the most remarkable properties of cycloids—namely, those connected with the families of tangents to these curves.

7.18. Prove that the tangents to a cardioid at the endpoints of a chord passing through the cusp of the cardioid are mutually perpendicular, and that their point of intersection is located at a distance $3r$ from the center of the stationary circle, where r is the radius of this circle. ↓

7.19*. Two pedestrians L and N walk at a constant speed around a circle. The ratio of their angular velocities is k (k is not 0, 1, or -1). Find the envelope of all the straight lines LN. ↓

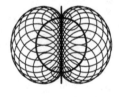

7.20*. Suppose we are given a circle and a straight line passing through its center. Prove that the union of all the circles whose centers lie on the given circle and which are tangent to the given straight line is a nephroid.

7.21*. Consider a Steiner deltoid drawn about a circle of radius $2r$ (the inscribed circle). Prove that an arbitrary tangent to the deltoid (at some point M) intersects the deltoid at two points K and L such that the segment KL has a constant length $4r$, and the midpoint of KL lies on the given inscribed circle. Prove also that the tangents to the deltoid at the points K and L are mutually perpendicular and intersect at a point N lying on the inscribed circle. Finally, show that the segments KN and LN are bisected by the inscribed circle. ↓

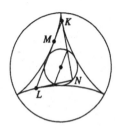

7.22*. Consider an astroid drawn about a circle of radius $2r$. Prove that from an arbitrary point of the inscribed circle P, it is possible to draw three straight lines PT_1, PT_2, PT_3, each tangent to the astroid, such that: (a) they form equal angles (of $60°$) with each other; and (b) the three points of tangency T_1, T_2, T_3 are the vertices of a right-angled triangle inscribed in a circle of radius $3r$, which is tangent to the circle described about the astroid.

The next and last problem in this series, which also may be solved using the language of motion, reveals an unexpected connection between the deltoid and the elementary geometry of a triangle. This curve is named after the geometer who discovered this connection.

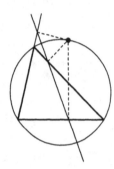

7.23*. A triangle ABC is given.

(a) Fix a point on the circumcircle of this triangle. Suppose we drop three perpendiculars from this point to each one of the triangle's sides. Prove that the feet of these three perpendiculars are collinear. (The line on which the three feet lie is called the *Wallace–Simson line* of the point on the circumcircle).

(b) Prove that the midpoints of the sides of a triangle, the feet of the altitudes and the midpoints of the segments of the altitudes joining the orthocenter to the vertices lie on a single circle (called the *nine-point circle*).

(c) Prove that all the Wallace–Simson lines of the triangle ABC are tangent to a single Steiner deltoid, drawn around the nine-point circle. ↓

Parametric equations

All the properties of cycloids may also be proved analytically, using coordinates. It is most convenient to write their equations in parametric form, expressing the coordinates (x, y) of the point M through a parameter l (the time). We have already come across these equations in Problem **6.22**.

Consider the locus of the fourth vertex M of a hinged parallelogram $OPMQ$ whose vertex O is at the origin in this coordinate system. (Note that $\overrightarrow{OM} = \overrightarrow{OP} + \overrightarrow{OQ}$). Suppose the point P moves with angular velocity ω_1 around the circle of radius r_1 whose center lies at the origin, O, of our coordinate system. Suppose the point Q moves with angular velocity ω_2 around the circle of radius r_2 whose center also lies at O. Then, at the moment t, the coordinates of P will be $(r_1 \cos \omega_1 t, r_1 \sin \omega_1 t)$; the coordinates of Q will be $(r_2 \cos \omega_2 t, r_2 \sin \omega_2 t)$; and the coordinates of the fourth vertex M of the parallelogram $OPMQ$ will be

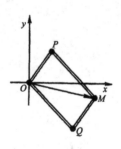

$$x = r_1 \cos \omega_1 t + r_2 \cos \omega_2 t$$
$$y = r_1 \sin \omega_1 t + r_2 \sin \omega_2 t.$$

(At the initial point of time $t = 0$, the sides OP and OQ of the hinged parallelogram are both directed along the axis Ox.)

In Problem **6.22** we saw that when $\omega_2 = -\omega_1$, the point M describes an ellipse. In the general case, when we have the following ratios:

$$\omega_1/\omega_2 = k, \qquad r_2/r_1 = |k|$$

the point M describes a k-cycloid.

When we eliminate t in the parametric equations, we obtain, in some cases, simple equations connecting the coordinates x and y. Consider, for example, the astroid. For this curve, we have $r_1 = 3r_2$, $\omega_2 = -3\omega_1$. We may take $\omega_1 = 1$. Then $\omega_2 = -3$, and the parametric equations of the astroid will be (putting $r_2 = r$):

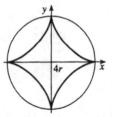

$$x = 3r \cos t + r \cos 3t$$
$$y = 3r \sin t - r \sin 3t$$

or more simply $\langle ? \rangle$:

$$x = 4r \cos^3 t, \qquad y = 4r \sin^3 t.$$

Hence we get the following equation of the astroid:

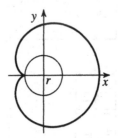

$$x^{2/3} + y^{2/3} = (4r)^{2/3}.$$

We can define the astroid and the other curves that we considered above by algebraic equations. Try to verify that the points (x, y) of these curves satisfy the following equations:

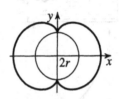

$$(x^2 + y^2 - 4r^2)^3 + 108r^2 x^2 y^2 = 0$$

107

(astroid)

$$(x^2 + y^2 - 2rx)^2 - 4r^2(x^2 + y^2) = 0$$

(cardioid)

$$(x^2 + y^2 - 4r^2)^3 - 108x^2r^4 = 0$$

(nephroid)

$$(x^2 + y^2 + 9r^2)^2 + 8rx(3y^2 - x^2) - 108r^4 = 0$$

(Steiner deltoid).

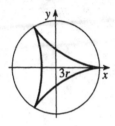

Thus, the astroid and the nephroid are curves of the sixth order, and the cardioid and Steiner deltoid are of the fourth order.

It can be proved that when $\frac{\omega_1}{\omega_2} = k$ is rational, cycloids are algebraic. When k is irrational, they are not; such curves pass arbitrarily close to any point of the ring that is centered at O and bounded by circles of radii $r_1 + r_2$ and $|r_1 - r_2|$. These curves are said to be "everywhere dense" in this ring.

Comparing the equations of the curves with their geometric properties yields new and interesting corollaries. Here is an example where a property of the astroid is used.

7.24. (a) Suppose we are given a right angle and, inside it, a point K whose distances to the two sides are a and b. Is it possible to draw through the point K a segment of length d whose endpoints lie on the sides of the right angle?

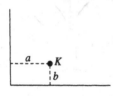

(b) A canal, whose banks are parallel straight lines, has a right angled turn in it. Before the turn, the width of the canal is a, and after the turn it is b. For what values of d can a thin log of length d pass around such a turn?

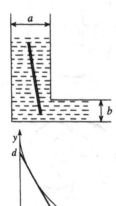

□ (a) Let us take the sides of the right angle as the coordinate axes. The segment of length d must touch an astroid whose cusps are at a distance d from the center. The equation of such an astroid is $x^{2/3} + y^{2/3} = d^{2/3}$. If the point K lies inside the region bounded by the astroid and the sides of the angle, then the required segment exists (it is a segment of the tangent to the astroid passing through the point K). If the point K lies outside this

108

region, it does not. Therefore, the necessary segment exists if and only if $a^{2/3} + b^{2/3} \leq d^{2/3}$. $\square$

Note that though we have found how to "construct," using an astroid, the required segment when the condition $a^{2/3} + b^{2/3} \leq d^{2/3}$ is satisfied, this problem cannot be solved using a ruler and compass.

Conclusion

The remarkable curves with which we have acquainted ourselves in the last two chapters have been known for more than two thousand years. The basic properties of ellipses, hyperbolas, and parabolas were described in the work *On Conics* by the ancient Greek mathematician Apollonius of Perga, who lived at about the same time as Euclid (third century B.C.). Even in ancient times, astronomers studied complicated circular motions. This is not surprising. If, in a very rough approximation, the planets are considered to be rotating around the Sun in circular orbits in a single plane, then the positions of another planet, as observed from the Earth, will be follow some kind of complicated circular motion. Over the centuries, however, astronomical observations grew increasingly refined, and the description of planetary motion by means of complicated cycloid curves underwent further and further modification. At long last, Johannes Kepler established that the planetary trajectories are ellipses with the sun located at one of the foci.

A wide range of problems from physics, mechanics, and mathematics were connected with particular curves. These provided a whetstone for sharpening the powerful analytical tools invented in the seventeenth century by Descartes, Leibniz, Newton, Fermat and others. These methods enabled the transition from particular problems connected with specific curves to general laws possessed by whole classes of curves. Needless to say, we cannot do without analytical methods when designing complicated mechanisms and constructions. However, the intuitive representations to which this book is devoted sometimes prove useful, even in problems not at all connected with geometry. It is not without reason that research or computational results are frequently represented in the form of graphs or families of lines.

CHAPTER 8

Drawings, Animation, and the Magic Triangle

What a pleasure it is to see or to draw nice geometric curves! Such curves can be drawn not only for fun, but also for very important practical or scientific reasons. Today one is not restricted to drawing with merely pencil and paper: one can write computer programs to produce graphs or go online to find sophisticated graphing capabilities on the Internet.

We will end this book by discussing how to obtain vivid images of geometric figures and then clarify the beautiful regularities that appeared in the last difficult problem (7.23*) of the book.

Certainly, drawings aid us in our understanding of the methods, ideas, and images that appear in our imaginations. But beware! They also can play tricks with our minds and may be the beginning of new illusions. For instance, the shape of an ellipse looks like an oval— a convex closed curve. But not every oval is an ellipse. In fact, it is not easy to recognize a real ellipse from among other ovals in a drawing. It is amazing indeed that the great Johannes Kepler was able to choose, from among the many different possible oval and circular orbits, the ellipse as the planetary trajectory (see p. 109).

The notion of a *locus of points* is also remarkably useful in defining geometric figures. Beautiful shapes such as circles, spheres, and ellipses can be expressed in simple statements. You may recall some of these definitions:

Circle/Sphere: In a plane (for a circle) or in space (for a sphere), the locus of points whose distances from a fixed point 0 (the center) are equal to a defined positive number r (the radius).

Ellipse: In a plane, the locus of points p with the property that the sum of the distances from p to two defined points A and B (the foci) is equal to a given positive number which is greater than the distance between the foci.

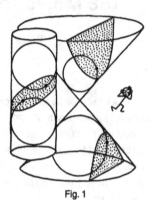

Fig. 1

Ellipses and other conics can be given as plane sections of a cone or cylinder. These figures can also be viewed as projections of a circle. The picture in Fig. 1 has been drawn by hand to illustrate this representation of an ellipse and a hyperbola. Try to imagine the four spheres in this picture: two inscribed in the cylinder and two inscribed in the cone which touch the corresponding sections at the foci of the ellipse and of the hyperbola. These spheres were invented by an engineer, Germinal P. Dandelin, to show the connection between the definitions of conics as plane sections and their definitions in terms of foci. These spheres are called *Dandelin spheres* in his honor. The frantic man in the picture, however, is drawn to illustrate our feelings about the process of creating three-dimensional drawings with computers.

Software engineers expend great effort creating computer models of three-dimensional figures useful in the manufacturing process. The design of these models usually starts with the simplest figures, often called

112

"primitives." In the 3-D drawing above, we see several primitives: a cylinder, a cone, spheres, and conics.

Drawing a family of curves is similar to drawing a picture in three dimensions. Recall that it can often be difficult to draw a real ellipse given its foci. We found (see Problem 6.5) a simple way to approximate, using a compass and straightedge, the family of ellipses with fixed foci A and B.

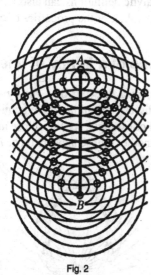

Fig. 2

a) Subdivide a segment AB into, say, 12 equal parts.

b) Draw 12 circles, each one of which has its center at A and passes through one of the other 12 points on the segment. This picture looks like a wave.

c) Analogously, draw 12 circles, each one of which has its center at B and passes through one of the other 12 points. These two families of circles together generate a net with 4-sided curved cells. (This picture may remind you of the interference of 2 waves).

d) Using small circles, mark the opposite vertices in a chain of the 4-sided curved cells.

We simultaneously obtain both the families of ellipses and hyperbolas with fixed foci. Using more subdivisions of AB will increase the smoothness of the curves connecting these vertices.

In the Introduction we found, using an equation, that a cat sitting on a ladder (away from the ladder's midpoint) moves along an ellipse as the ladder slides toward the floor.

This was our first example of the process of converting geometric definitions into analytic formulas. In fact, the analytic formula for a curve describes the curve as a locus of points whose coordinates satisfy the formula. Analytic formulas can be input into computer programs to create very accurate geometric shapes. Convenient analytic definitions can also quickly bring us to an understanding of the properties of curves.

The envelope of a family of lines

Sometimes, in a drawing of a family of straight lines, we can discern a familiar curve bounded by the straight lines. For example, see the ellipse in the drawing in Fig. 3 (another ellipse appears on p. 76).

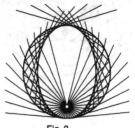

Fig. 3

This drawing was done by the following construction:

a) Draw a circle with center O and choose a point A inside of the circle;

b) Subdivide the circle in, say, 24 equal parts;

c) Through each point M of this subdivision, draw the straight line that is perpendicular to the segment MA.

In such constructions, the fixed point A is called the *pedal*.

Thus, if the point M is moving along the circle, then the perpendiculars to the segments MA are tangents to the ellipse. In other words, the ellipse is the envelope of the one-parameter family of the constructed straight lines.

114

We also saw another appearance of a curve as an envelope of a family of lines in Problem **7.16**; this was called a *caustic curve*. When light reflects off of a curve, the envelope of the reflected rays is a *caustic by reflection*.

Let us consider a small piece of a smooth curve C that does not contain any straight line segments. We say that the curve C is the envelope of the one-parameter family of its tangent lines. Thus, if a point M moves along the curve, then the tangent line to M moves as well. The trajectory of a moving point M is the envelope of the family of straight lines generated by the velocity vector of the moving point M.

For instance, recall the definition of a Steiner deltoid as a (-2)-cycloid (see Problem **7.13**): Suppose that inside a stationary circle with radius R, we have another circle whose radius is $2R/3$. Next, suppose this smaller circle is internally tangent to the larger circle and rolls along the larger circle without sliding. Steiner's deltoid is given by the trajectory of a point on the smaller circle.

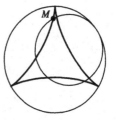

It turns out that it also can be defined as an envelope of a family of straight lines—just in a slightly different way. The family of tangent lines to the Steiner deltoid can also be represented as positions of the moving diameter of a moving circle. Imagine that inside the same stationary circle with radius R is a circle whose radius is $2R/3$. Suppose this circle also touches the same larger circle from the interior and rolls around the larger circle without sliding. Let PQ be a fixed diameter of this circle of radius $2R/3$ (again, see Problem **7.13**). The envelope of all positions of this diameter PQ yields the Steiner deltoid. Such a curious representation of the Steiner deltoid is a consequence of the theorem on two circles (see p. 99 in Chapter 7). Later on, we will see yet another realization of the Steiner deltoid as the envelope of a family of lines, defined in yet another way.

The magic triangle

We know a triangle can be "decorated" by its circumscribed circle, inscribed circle, their centers, the or-

thocenter and the centroid. New constructions or decorations we can add to a triangle include the Feuerbach
Circle, the Steiner deltoid of Wallace–Simson lines,
and Morley's triangle. An ordinary triangle with these
wonderful decorations, particularly one in which these
points and lines move around, is a figure we will call a
Magic Triangle.

An equilateral triangle is a beautiful, simple primitive in a plane. Curiously enough, any triangle has
certain equilateral triangles that are closely connected
to it. First of all, we will discuss the Steiner deltoid
(Property (c) below) which leads to Steiner's equilateral triangle, and then, at the end of the chapter, we will
discuss Morley's triangle.

Remember that the **circumcircle** of a triangle is the
circumscribed circle of the triangle, and the **orthocenter** of a triangle is the point of intersection of its three
altitudes.

Any triangle ABC (Problem **7.23***) has the following fascinating properties:

a) The nine-point circle. The three midpoints of the
sides of a triangle, the three feet of altitudes, and the
three midpoints of segments of the altitudes joining the
orthocenter to the vertices lie on a single circle. This
circle is called the *Nine-point Circle* or the *Feuerbach
Circle*.

b) Let M be any point on the circumscribed circle
of the triangle. The feet of the perpendiculars dropped
from M to the sides AB, BC and AC (or their extensions) all lie on a single line called the *Wallace–Simson*
line.

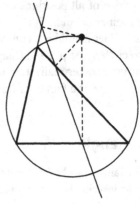

Fig. 4

c) When the point M moves around the circum-scribed circle, the envelope of the resulting family of Wallace–Simson lines is a **Steiner deltoid**. This Steiner deltoid itself is tangent at three points to the Feuerbach circle of the triangle.

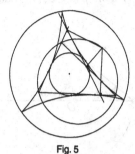

Fig. 5

We shall explain these connections by considering points and lines in motion. Recall the animated basic images from the beginning of this book.

A tiny ring on a circle

Notions of an angle and its measure can be represented by a clock and the rotations of its hands. Usually, counterclockwise rotation is called *positive orientation* and clockwise rotation is *negative*. We will use this convention throughout.

Now, recall the familiar situation in Chapter 1: suppose a small ring is put on a wire circle. A rod passing through this ring rotates around the point A of the circle. We discovered (see p. 11) that the angular velocities of the moving rod and the tiny ring are different:

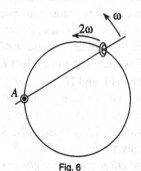

Fig. 6

117

If the rod rotates uniformly with an angular velocity ω, the ring also moves around the circle uniformly but with angular velocity of 2ω, that is, twice the velocity of the rod. This regularity is closely related to the theorem *concerning the inscribed angle in a circle* and *Copernicus' Theorem* (see the Introduction).

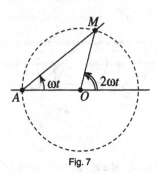

Fig. 7

Two pedestrians on a circle and the Steiner deltoid

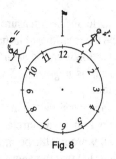

Fig. 8

We have already seen that many different definitions can be used to describe the same curve. Naturally, we should try to choose the most convenient definition for our particular goal. In this case, for our definition of Steiner's deltoid, we will use the notion of an envelope of a family of lines as given in Problem **7.19**.

Two pedestrians L and N (see 7.19*) walk at constant speeds around a circle in opposite directions: counterclockwise and clockwise, respectively. But the pedestrian L walks twice as fast as N. Suppose their angular velocities are equal to 2ω and $(-\omega)$, respectively. Then the *envelope of straight lines LN will be* *a Steiner deltoid.* We can see this from the following experiment:

Consider the face of a round clock (a circle) divided into 12 equal parts, marked with the numbers from 1 to 12. Suppose that at the initial moment the pedestrians L and N are on the number 12. Let us represent this by the ordered pair (12,12) and let us draw the horizontal tangent line to the circular clock at the point 12. When N takes one step to the number 1, L takes two steps in the opposite direction to the number 10. At this point, their positions are given by the pair (10,1). So let us draw a line passing through both 10 and 1.

As the pedestrians continue their movement, we obtain the following sequence of ordered pairs for their positions:

$(12,12) \longrightarrow (10,1) \longrightarrow (8,2) \longrightarrow (6,3) \longrightarrow (4,4)$
$\longrightarrow (2,5) \longrightarrow (12,6) \longrightarrow (10,7) \longrightarrow (8,8) \longrightarrow (6,9) \longrightarrow$
$(4,10) \longrightarrow (2,11) \longrightarrow (12,12).$

For each ordered pair whose entries are distinct, such as (2,11) and (6,9), we draw a line connecting the corresponding positions of the pedestrians on the circular clock. The pedestrians will meet each other three times–at the points 12, 4, and 8 on the clock–and at each such point where they meet, we draw tangent lines to the circle. These three meeting points are the vertices of an equilateral triangle. When we draw the lines that connect the sequential positions of pedestrians, we find that the *envelope of these lines is a Steiner deltoid.*

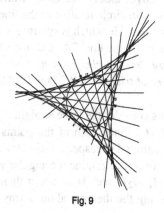

Fig. 9

For pictorial clarity, it is better to subdivide the circle not into 12 but into 24 parts. The picture will be clearer still if, through each line LN connecting the pedestrians, we draw segments of lines with equal lengths and with their midpoints at the points N. Later on, we will need the value of the angular velocity of the line LN.

From this experiment, we find that if the pedestrian N moves clockwise with angular velocity $(-\omega)$, the line LN rotates counter counterclockwise twice as slowly, e.g., with angular velocity $\omega/2$. Indeed, take any point O on a plane and after every step of the pedestrian N, draw the line through the point O parallel to LN.

How three points move around three symmetric circles

Let us take a point M on the circumcircle of the triangle ABC, and let us suppose that it moves around the circle clockwise with angular velocity $(-\omega)$. See Fig. 10(a) below.

Let the points M_1, M_2 and M_3 be symmetric to the point M relative to the *lines BC, CA and AB*. (By *lines* we mean the sides and the extensions of the sides AB, BC, and CA of the triangle ABC.) Since the point M is moving clockwise with angular velocity $(-\omega)$, each of the points M_1, M_2 and M_3 moves around the circle which is symmetric to the circumcircle relative to the lines BC, CA and AB, respectively—that is, counterclockwise—with angular velocity ω. (So, for instance, M_1 moves around the circle which is symmetric to the circumcircle relative to the line BC; M_2 moves around the circle which is symmetric to the circumcircle relative to the line CA, and so on.) We have seen that *these three circles meet at a one point H, the orthocenter of triangle ABC* (*see Problem* **3.8** *on* p. 39).

Now, let us consider the three straight lines M_1H, M_2H and M_3H. Since each of the points M_1, M_2, and M_3 moves around its respective circle with angular velocity (ω), we can determine the angular velocity of the lines M_1H, M_2H and M_3H about their common point H by using the theorem about a tiny ring on a

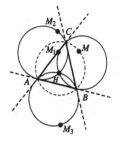

Fig. 10(a)

circle. Indeed, as a consequence of the theorem, these three straight lines all rotate with angular velocity $\omega/2$ about their common point H.

Let us note, though, that M_1H, M_2H and M_3H are not three different straight lines: rather, they all coincide, giving us a single line. While this may seem intuitively clear from the diagram, keep in mind that Fig. 10(a) only displays one *particular* configuration of the points M, M_1, M_2, and M_3. To prove that the lines M_1H, M_2H and M_3H always coincide, however, we must show that these lines coincide for *every* position of M as M moves around the circle.

The key step is to recall that the three lines all rotate with the same angular velocity about H, so if the three lines coincide at one moment in time, then they coincide everywhere. Now, consider the moment in time when the point M is at at the vertex C of the triangle. Then M_1 and M_2 are the same point, so the lines M_1H and M_2H coincide at that instant, and hence these two lines coincide everywhere. Similarly, consider the moment in time when M is at the vertex B of the triangle. Here, the points M_1 and M_3 are the same point, so the lines M_1H and M_3H coincide everywhere. Therefore, these three lines are identical, and at each moment of rotation, the three points M_1, M_2, M_3 belong to a single straight line l passing through H. See Fig. 10(b).

To get an indication of what happens as M moves, examine Fig. 10(c). Let's start again with the triangle and its circumcircle, on which the point M rotates clockwise. This is indicated by the clockwise arrow on the circumcircle. Draw the three circles that are symmetric to the circumcircle relative to the three sides of the triangle. We know that all three circles intersect at the orthocenter H of the triangle. As M moves, the points M_1, M_2, and M_3 rotate counterclockwise on their respective circles. Note, however, that unlike Figs. 10(a) and 10(b), Fig. 10(c) does *not* indicate a specific configuration of the points M, M_1, M_2, and M_3–rather, it demonstrates the *directions* and the circles traversed by the points. We summarize our results as follows:

There exists a straight line l which rotates counterclockwise about the point H with angular velocity $\omega/2$

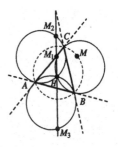

Fig. 10(b)

and is connected to the moving point M in the following way:

The three points M_1, M_2, and M_3 are actually the moving points of intersection between this rotating line and each of the three circles. Each of these points rotates about its circle with angular velocity equal to ω, and each of these points lies on lies on the line l.

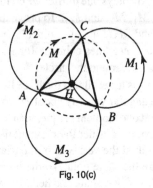

Fig. 10(c)

Finally, we will describe a wonderful moving picture that illuminates the Wallace–Simson line.

The Wallace–Simson line

As above, let M be a point on the circumcircle of the triangle ABC. The feet of the three perpendiculars from M to the lines AB, BC, and AC all lie on a single line. Indeed, if we extend each of these three perpendiculars to twice its original length, we will get three points M_1, M_2, and M_3 which are symmetric to the point M relative to the three sides (or the extensions of the three sides) of the triangle. But we know that these three points lie on a single line l. In other words, let us fix a position of the point M; then the Wallace-Simson line is the locus of midpoints of all segments ML, where L is any point of the line l.

All of the Wallace-Simson lines of the triangle ABC touch a Steiner deltoid. That is, the envelope of Wallace-Simson lines is a Steiner deltoid. For an explanation of this relationship between the Wallace-Simson lines and the Steiner deltoid, we will start by discussing the aforementioned nine-point or Feuerbach circle.

122

The Nine-Point Circle

Suppose we are given a triangle ABC and its circumcircle. Let O denote the center of the circumcircle. Suppose, as before, that a point M on the circumcircle rotates clockwise with angular velocity $-\omega$. Let H be the orthocenter of the triangle, and let K be the midpoint of the segment MH. As M moves around the circumcircle, the point K moves around a smaller circle. This smaller circle is similar to the circumcircle, with ratio of similitude $1/2$ and center of similitude H (see Problem **3.20**, p. 43). Its center is the midpoint O_1 of the segment connecting 0 and H. Indeed, you can easily verify that as K moves, the length of the segment $O_1 K$ remains constant and is equal to half the radius of the circumcircle.

The circle traversed by K passes through nine particular points of interest in the triangle ABC, and hence it is known as the *nine-point circle*. It is also called the *Feuerbach* or *Euler* circle. The nine points in the triangle are the following (see Fig. 10(d)):

a) The three midpoints of the segments joining the orthocenter H to the vertices A, B, and C;

b) The three midpoints of the sides of the triangle; and

c) The three feet of the altitudes of the triangle.

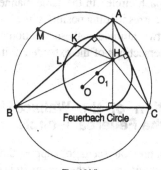

Fig. 10(d)

To prove this, we will show that for each of the nine points given above, we can exhibit M on the circumcircle so that the given point is the midpoint of HM.

First, as M moves around the circumcircle, it moves through the vertices A, B and C, so the point K on the Feuerbach circle moves through the midpoints of segments HA, HB and HC.

Next, consider the arc of the circumcircle that is cut off by the chord AB. We know that if we reflect this arc across the line AB, we will get a symmetric arc that passes through the orthocenter H. Let L be the midpoint of AB. Suppose we extend the segment HL to twice its length; by symmetry, this extended segment meets the circumcircle. By construction, this position of M on the circumcircle is such that L is the midpoint between H and M. Hence L lies on the Feuerbach circle. Analogously, we can show that the midpoints of the other two sides, BC and CA, also lie on the Feuerbach circle.

Finally, we will demonstrate that the feet of the three altitudes of the triangle lie on the Feuerbach circle as well. Again, suppose we construct the arc symmetric to the circumcircle relative to the segment AB. This arc passes through the orthocenter H. Consider the altitude whose foot lies on AB. Recall that the altitudes all pass through H as well. If we extend this altitude until it intersects the circumcircle, then, by symmetry, we obtain a position of M on the circumcircle for which the segment HM is bisected by the foot of the altitude. That is, the foot of the altitude is the midpoint of the segment HM, and hence the foot of the altitude lies on the Feuerbach circle. In the same manner, we can construct the corresponding positions of M for the feet of the other two altitudes—these positions are, once again, the intersections of the extended altitudes with the circumcircle.

The rotation of the Wallace–Simson line and the Feuerbach circle

Recall our previous scenario: suppose that the point M moves clockwise around the circumcircle of triangle ABC with angular velocity $(-\omega)$. The midpoint M_{mid} of the segment HM, where H is the orthocenter of the ABC, moves with the same angular velocity $(-\omega)$.

The Wallace–Simson line rotates with angular velocity $\omega/2$, because it is always parallel to the line $M_1 H$, where M_1 is the point symmetric to M relative to the side BC of the triangle ABC (as we have seen, the two other points M_2 and M_3 also lie on M_1H). Also, the Wallace–Simson line goes through M_{mid}. Thus, the other point of intersection of the Wallace–Simson line with the nine-point circle rotates with angular velocity 2ω.

These intersection points of the Wallace–Simson line move around a circle like the two pedestrians L and N we encountered before: the point M_{mid} moves like the pedestrian N, with angular velocity $(-\omega)$, and the other intersection point S moves like the pedestrian L, with angular velocity 2ω. Thus the envelope of lines joining their positions M_{mid} and S is a Steiner deltoid. Therefore, the family of tangent lines to a Steiner deltoid has three representations: as the moving diameter of a moving circle; as the moving line connecting two moving pedestrians; and as the family of Wallace-Simson lines in a triangle.

Steiner's triangle and Morley's triangle

The three vertices (cusps) of the Steiner deltoid generate an equilateral triangle. Let us call this equilateral triangle *Steiner's triangle*.

Surprisingly, there exists another equilateral triangle that also can be derived from any triangle.

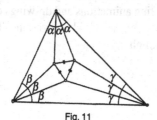

Fig. 11

Consider an arbitrary triangle ABC. Suppose we trisect each of its angles. Recall that to trisect an angle means to divide it evenly into three equal angles. In trisecting each angle, we construct two rays that partition the angle into three equal, smaller angles. Now, for each side of the triangle, consider the four rays we have constructed from the two angles adjacent to this side. Choose the two rays (one from each angle) that are closer to each side and consider their point of intersection. We obtain three such intersection points, one for each side of the triangle. It turns out that these three points are always the vertices of an equilateral triangle. This triangle is called *Morley's triangle*.

In our picture, it appears that Morley's triangle and Steiner's triangle have parallel sides.

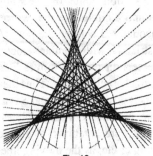

Fig. 12

It might be worthwhile for to repeat our experiments to verify these regularities. Also, elegant demonstrations of these facts can be found on the Internet: for instance, we located an interesting paper written by Miguel De Guzman with a nice proof that Morley's triangle and Steiner's triangle have parallel sides. There are a number of websites that feature unusual drawings and discussions of plane geometry, and the Internet is replete with different types of graphing software. Indeed, fascinating animations and drawings of lines and curves are everywhere, vivid realizations of geometric beauty and simplicity.

Answers, Hints, Solutions

1.13. Note that the vertices M of the right triangle AMB with hypotenuse AB lie on the circle with the diameter AB.

1.14. Let us draw the common tangent through the point of contact M of the circles. Let it cross AB at the point O. Then $|AO| = |OB| = |OM|$ (the lengths of the tangents from the point O to the circles are equal).

1.15. *Answer*: The union of three circles. Let A, B, C and D be the given points. Draw a straight line l through the point A, a line parallel to l through the point C, and straight lines perpendicular to l through the points B and D. The result is a rectangle.

Let L be the midpoint of the segment AC and K the midpoint of the segment BD. Then it is easy to see that $\widehat{LMK} = 90°$, where M is the center of the rectangle. Rotating l about the point A and rotating the other straight lines correspondingly, we find that the set of the centers M of the constructed rectangles is a circle with the diameter KL.

Now the four points A, B, C, D may be divided into two pairs in three different ways: (A, C) and (B, D); (A, B) and (C, D); and (A, D) and (B, C). Therefore, the required set consists of three circles.

1.25. *Answer*: Either the midpoint remains fixed or it moves along a straight line. If the pedestrians P and Q move along parallel straight lines with velocities that are equal in magnitude and opposite in direction, then, clearly, the midpoint of the segment PQ remains stationary; the path of the midpoint in this case is simply a single fixed point. If the pedestrians move along parallel straight and their velocities are *not* equal in magnitude and opposite in direction, then the midpoint of the segment PQ also moves along another parallel straight line.

To see why, suppose the straight lines intersect at the point O. Regard O as the origin. Then the velocities $\vec{v_1}$ and $\vec{v_2}$ of the pedestrians are vectors directed along straight lines, and their values are equal to the lengths of the paths walked by the pedestrians in unit time. Let the first pedestrian be situated at the point P at time t, and the second one at the point Q. Then $\overrightarrow{OP} = \vec{a} + t\vec{v_1}$ and $\overrightarrow{OQ} = \vec{b} + t\vec{v_2}$ (where the vectors $\vec{a}$ and $\vec{b}$ define the initial positions of the pedestrians when $t = 0$).

The midpoint of the segment PQ is at the point M where

$$\overrightarrow{OM} = \frac{\overrightarrow{OP} + \overrightarrow{OQ}}{2} = \frac{\vec{a} + \vec{b}}{2} + t\frac{\vec{v_1} + \vec{v_2}}{2}.$$

So the point also moves along some straight line with a constant velocity $\frac{\vec{v_1} + \vec{v_2}}{2}$. In order to find this line, it is sufficient to mark the midpoint of the initial positions of the pedestrians and their positions after one unit of time.

We may replace the vector calculations by the following geometric argument.

If P_0P_1 and Q_0Q_1 are any two nonparallel segments, then the segment M_0M_1, where M_0 and M_1 are the midpoints of the segments P_0Q_0 and P_1Q_1, is a median of the triangle $L_1M_0N_1$, where L_1 and N_1 are the fourth vertices of the parallelograms $P_1P_0M_0L_1$ and $Q_1Q_0M_0N_1$. (See Fig. 1. In the construction depicted, $P_1L_1Q_1N_1$ is a parallelogram, and P_1Q_1 and N_1L_1 are its diagonals.)

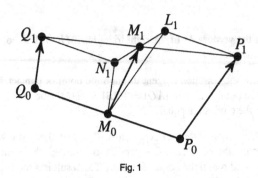

Fig. 1

It is now clear that if, instead of P_1 and Q_1, we take points on the lines Q_0Q_1 and P_0P_1 such that $\overrightarrow{P_0P} = t\overrightarrow{P_0P_1}$ and $\overrightarrow{Q_0Q} = t\overrightarrow{Q_0Q_1}$, and the corresponding triangle LM_0N (with the median M_0M) is drawn, as before, then this triangle may be obtained simply by a similarity transformation with coefficient t and center M_0 from the triangle $N_1M_0L_1$ (with the median M_0M_1), i.e., the point M will lie on the straight line M_0M_1 and $\overrightarrow{M_0M} = t\overrightarrow{M_0M_1}$.

1.28. Let us use Fig. 1 to Problem **1.25**. If the segments P_0P_1 and Q_0Q_1 rotate uniformly about the points P_0 and Q_0 with equal angular velocities (at one revolution per hour), then the triangle $N_1M_0L_1$ also rotates, along with its median M_0, as a rigid body about the point M_0 with the same angular velocity.

1.29. *Answer*: A circle. Let us translate this problem into the language of motion. Draw the radii O_1K and O_2L. Let the straight line KL rotate with a constant angular velocity ω.

Then according to the theorem about the tiny ring on a circle, the radii O_1K and O_2L will rotate uniformly with the same angular velocity 2ω, i.e., the size of the angle between the radii O_1K and O_2L remains constant. Thus the problem reduces to the previous one.

2.11. (b) Use Proposition F.

2.19. *Answer*: If h is the height of triangle ABC, then the required set is empty when $\mu < h$, the entire triangle (Fig. 2) when $\mu = h$, the contour of a hexagon (Fig. 3) when $\mu > h$.

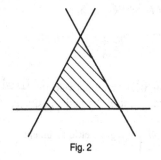

Fig. 2

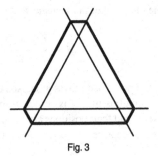

Fig. 3

128

2.20. (b) See Fig. 4.

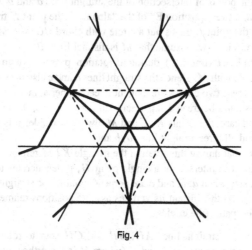

Fig. 4

3.5. (b) The problem reduces to **3.5 (a)** and is simply solved by "embedding in space": If three spheres are constructed with their centers in the plane α on the given circles (in the horizontal plane α) and looked at from above, then we see three circles in which the spheres intersect (their projections on the horizontal plane are our three chords) and also their point of intersection (its projection is the required point of intersection of the chords).

3.7. (b) Note that $\widehat{AMB} = 90° + \frac{\varphi}{2}$, where M is the center of the inscribed circle of the triangle. According to **E** the set of points M is a pair of arcs together with their endpoints A, B.

3.7. (c) *Answer*: The required set is a pair of arcs (see Fig. 5 a, b, c, for each of the corresponding cases: (a) $\varphi < 90°$, (b) $\varphi = 90°$, (c) $\varphi > 90°$).

Let l_A and l_B be two intersecting straight lines passing through the points A and B, respectively, and let k_A and k_B be the straight lines passing also through A and B, such that $k_A \perp l_B$, $k_B \perp l_A$. If the lines l_A and l_B rotate about their points A and B, then k_A and k_B also rotate about their points A and B with the same constant angular velocity. According to Proposition $\mathbf{E°}$, the point of intersection of k_A and k_B moves in a circle.

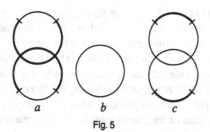

a b c

Fig. 5

Note that when the point of intersection of the straight lines l_A and l_B moves along an arc of the circle γ, the point of intersection of the straight lines k_A and k_B also moves along an arc of a circle; it is symmetric to the circle γ relative to the straight line AB.

129

3.8. (a) Let a, b, c be straight lines passing through the points A, B, C, respectively. Let K, L and M be the points of intersection of the straight lines a and b, b and c, a and c, respectively. According to Proposition $\mathbf{E}°$ of the "alphabet" the point K traces out a circle with chord AB, and the point L traces out a circle with chord BC. Let M be the point of intersection of these circles—we assume that M is distinct from B.

When the straight line b (line KL), during its rotation, passes through the point H, the points K and L coincide with M. Hence the straight lines a and c also pass through H. (The particular case when these two circles are tangent to each other at the point B and the case when they coincide should be treated separately. In the first case, the point M coincides with B. In the second case, the points K, L and M always coincide: it is possible to put a single tiny ring around all three straight lines a, b and c.

Incidentally, note that during this rotation the triangle KLM remains similar to itself. When all the straight lines intersect at a single point H, it degenerates to a point, and it attains its maximum size when a, b, and c are perpendicular to the straight lines AH, BH, and CH, respectively. At this instant its vertices assume positions diametrically opposite to the point H on their paths (the circles).

3.8. (b) Suppose the straight lines AH, BH and CH start to rotate with the same angular velocity about the points A, B and C (where H is the orthocenter of the triangle ABC). Then the point of intersection of each pair of straight lines describes one of the circles mentioned in the statement of the problem.

3.9. Consider three sets of points M lying inside the triangle:

$$\left\{ M : \frac{S_{AMB}}{S_{BMC}} = k_1 \right\}, \qquad \left\{ M : \frac{S_{BMC}}{S_{AMC}} = k_2 \right\}, \qquad \left\{ M : \frac{S_{AMC}}{S_{AMB}} = k_3 \right\}.$$

These three segments (see Proposition **I**) are concurrent when and only when $k_1 k_2 k_3 = 1$.

3.10. Consider three sets:

$$\{ M : |MA|^2 - |MB|^2 = h_1 \}, \qquad \{ M : |MB|^2 - |MC|^2 = h_2 \},$$
$$\{ M : |MC|^2 - |MA|^2 = h_3 \}.$$

These three straight lines (see Proposition **F**) are concurrent if and only if $h_1 + h_2 + h_3 = 0$.

3.18*. (a) We need to prove that we can find a point in the forest that is at a distance greater than A/P from the edge of the forest. Let us work by contradiction: suppose every point inside the polygon lies at a distance of A/P (or less) from the edge of the forest.

From each side of the polygon, then let us construct rectangles such that each rectangle has width A/P and overlaps the polygon. (That is, one side of the rectangle coincides with one side of the polygon, and the rectangle has width A/P). If our assumption is true, then all of these rectangles will fully cover the polygon. Also, every pair of neighboring rectangles will overlap. Hence the sum of their areas must be strictly greater than the area A of the polygon.

On the other hand, if $a_1, a_2, a_3, \ldots, a_n$ are the lengths of sides of the polygon, then the sum of the areas of the rectangles we constructed is equal to $a_1(A/P) + a_2(A/P) + a_3(A/P) + \cdots + a_n(A/P) = (a_1 + a_2 + a_3 + \cdots + a_n)A/P = P(A/P) = A$.

130

This leads us to the contradiction that $A > A$. Therefore, our initial assumption—namely, that there was no point in the interior that was at a distance of more than A/P from the polygon's sides—must be false.

Now, a *convex polygon* is the intersection of a set of half-planes, each bounded by the line on one of its sides. In the solution above, we implicitly used the following properties about convex polygons: First, in order for a point to lie inside the convex polygon, it has to lie in each half-plane. Because of this, we could construct rectangles on each side of the polygon which overlap with the polygon.

Also, any interior angle of a convex polygon is always less than 180 degrees. As a result, every pair of adjacent rectangles must overlap with one another.

3.21. Draw the set of endpoints M of all the possible vectors

$$\overrightarrow{OM} = \overrightarrow{OE}_1 + \overrightarrow{OE}_2 + \cdots + \overrightarrow{OE}_n$$

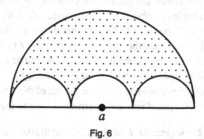

a

Fig. 6

(where $\overrightarrow{OE}_i$ are the unit vectors mentioned in the statement of the problem, first for $n = 1$, then for $n = 2$, and so on (Fig. 6).

4.4. *Answer*: The minimum distance between the pedestrians is equal to $u/\sqrt{u^2 + v^2}$.

Suppose the first pedestrian P walks with velocity $\overrightarrow{u}$, the second, Q, with velocity $\overrightarrow{v}$ (the lengths of u and v of these vectors are known). Consider the relative motion of P in the reference frame of the pedestrian Q. This will be a uniform motion with a constant velocity $\overrightarrow{u} - \overrightarrow{v}$ (see **1.3**).

In the "initial" position, when P lies at the point P_0 where the roads cross, Q_0 is a distance $|Q_0 P_0| = d$ from P_0 in the direction of the vector $-v$. Thus, in order to find the answer, it is sufficient to draw through the point P_0 a straight line l parallel to the vector $\overrightarrow{u} - \overrightarrow{v}$ (it is the path of P in its relative motion in the reference frame of Q), and to determine the distance $|Q_0 H|$ of the point Q_0 from the straight line l (H is the projection of Q_0 on l). Since the triangle $Q_0 P_0 H$ is similar to the triangle formed by the vectors $\overrightarrow{u}$, $\overrightarrow{v}$ and $\overrightarrow{u} - \overrightarrow{v}$ $((Q_0 P_0) \perp \overrightarrow{u}, (Q_0 H) \perp (\overrightarrow{u} - \overrightarrow{v}))$, we have

$$|Q_0 H|/|Q_0 P_0| = |\overrightarrow{u}|/|\overrightarrow{u} - \overrightarrow{v}| = u/\sqrt{u^2 + v^2}.$$

4.6. From the center O_1 of one of the circles, drop a perpendicular $O_1 N$ onto the secant l passing through the point A, and from the center O_2 of the other circle, drop a perpendicular $O_2 M$ onto the straight line $O_1 N$. Then the length of $|O_2 M|$ is half the distance between the points of intersection of the secant l and the circles (other than A).

4.9. *Answer*: An isosceles triangle. Use **2.8** (a).

4.12*. (a) *Answer*: The crocodiles should be at the midpoints of the sides of an equilateral triangle inscribed in the circle of the lake. To see why, suppose the lake has radius R. Suppose crocodiles are placed so that the distance from each point of the lake to the closest crocodile is no greater than some fixed value, say r; in other words, the three circles—each centered at a particular crocodile, and each with radius r—cover the entire lake. Note that these three circles must contain the circumference of the lake. Therefore, each of the circles overlaps with an arc of the lake's circular boundary. Consequently, one of these arcs must subtend an angle of at least 120 degrees (and less than 180 degrees). Recall that this arc is an arc of the lake's circular boundary; the lake's circle has radius R. The maximum possible distance between two points of this arc is greater than the length of the chord of the 120° arc: that is, greater than $a = R\sqrt{3}$. But also recall that this arc is covered by a circle of diameter $2r$, and thus $2r$ must be greater than $R\sqrt{3}$; this means that r must be greater than $\frac{R\sqrt{3}}{2}$.

So, no matter where crocodiles are placed, the value of r is no less than the half of the side a of an equilateral triangle inscribed in the circle with radius R. Now, if the crocodiles are placed at the midpoints of the sides of an equilateral triangle inscribed in the circle of the lake, then the value of r is $a/2$ and the three circles of radius r centered at each crocodile cover the entire lake. Hence the arrangement is optimal. A similar argument works for four crocodiles. *You may wish to explore further: What if there are five crocodiles?*

(b) *Answer*: The crocodiles should be at the midpoints of the sides of a square inscribed into the circle of the lake. The proof uses an argument similar to that for part (a).

5.4. (b) Prove that if the segment KL of constant length slides at its endpoints along the sides of the given angle A, then the point M of intersection of the perpendiculars erected at the points K and L to the sides KA and LA of the angle moves around a circle with center A (recall the discussion of Copernicus' Theorem **0.3** in the Introduction).

5.7. The following fact helps us construct these points: the level curves of the function $f(M) = |AM|/|MB|$ are orthogonal to the circles passing through the points A and B (p. 97).

6.3. (c) *Answer*: A hyperbola, if the given circles do not intersect one another (or are tangent); the union of a hyperbola and an ellipse, if they intersect; an ellipse, if one circle lies inside the other (or are tangent). The foci of the curves lie at the centers of the given circles.

In order to reduce the number of different cases that must be considered, we can use the following general rule: circles of radii r and R with their centers a distance d apart, touch each other if $r + R = d$ or $|R - r| = d$.

6.12. (a) For a given tangent, construct the tangent symmetric to it with respect to the center of the ellipse.

Use **6.9** (b) and the theorem which states that the product of the segments of a chord drawn through a given point inside a circle is independent of the direction of the chord.

6.15. In case (a), construct an ellipse (and in the case (b), a hyperbola) with foci at A and B, touching the first link P_0P_1, and prove that it also touches the second link P_1P_2. To do this use the fact that $\triangle A'P_1B \cong \triangle AP_1B'$, where A' is the point symmetric to A

relative to $P_0 P_1$ and B' the point symmetric to B relative to $P_1 P_2$. The tangents will be the perpendicular bisectors of the segments AA' and BB' (**6.9** (a), **6.10** (a)).

6.16. (c) We construct the set of points N for which the midpoint of the segment AN lies on the given circle. This is another circle. Denote its center by B and its radius by R. The set of points which are located nearer to the point A than to any point N of the constructed circle is the intersection of the half planes containing A which are bounded by the perpendicular bisectors of the segment AN. This set may be written as follows:

$$\{M : |MA| - |MB| \leq R\},$$

i.e., its boundary is a branch of a hyperbola.

6.17. Compare the hint to **6.16** with the proof of the focal property of a parabola.

6.23. Choose the origin at the midpoint of the segment AB, and the x-axis so that at some points of time both rotating straight lines are parallel to Ox. If we write the equations of the straight lines for time t, find the coordinates of their point of intersection and then eliminate t (as in the solution to **6.22**), then we obtain the equation of a hyperbola in the form (4) (p. 80).

6.24. Imagine two straight lines rotating about the points A and B in different directions so that the second one has twice the angular velocity of the first. It is not difficult to guess that their point of intersection moves along a curve that looks like a hyperbola whose asymptotes form angles of $60°$ with the straight line AB and for which the point of intersection C with the segment AB divides AB in the ratio $|AC|/|BC| = 2$.

The answer to this problem is, in fact, a branch of a hyperbola. The following simple geometrical proof reduces the problem to Proposition N of the "alphabet."

Construct the point M' symmetric to M with respect to the midperpendicular l of the segment AB, and note that the ray BM' is the line bisector of the angle ABM, and $|MM'| = |MB|$, so that $|MB|/\rho(M, l) = 2$.

6.25. (a) If the coordinate system is selected in such a way that the sides of the angle are given by the equations $y = kx$ and $y = -kx, x \geq 0$, then the area of the triangle OPQ, where P and Q lie on the sides of the angle is $kx^2 - y^2/k$, where $(x; y)$ are the coordinates of the midpoint of the segment PQ.

(b) Use the result of Problem **1.7** (b).

(c) Result follows from (a) and (b).

7.2. This union may be considered as the set of those points M for each of which there can be found a point P on the circle such that $|MP| \leq |PA|$ or as the set of points M for which the perpendicular bisector of the segment MA has a point in common with the circle. Compare this problem with **6.16–6.17**.

7.9. *Answer:* (a) 3; (b) 4; (c) 2.5. The ratio of the angular velocity may be found in the same way as was done in the examples on pp. 101–103.

7.13. (a) The arc of a circle of radius R between two cusps of a Steiner deltoid ($120°$) has the same length as the circumference of a semicircle of radius $2R/3$.

7.14. (b) Both curves may be obtained as the paths of the vertex M of a hinged parallelogram, with side lengths $R - r$ and r. The ratio of the angular velocities ω_1/ω_2 is equal to $-r/(R - r)$ (the angular velocities have opposite signs; see p. 101).

7.18. Use **7.7** and Mozzi's theorem.

7.19. *Answer*: a k-cycloid (see p. 96).

7.21. Use **7.13** (a), Mozzi's theorem and the theorem on two circles.

7.23. Let M be a point on the circle described, moving around it with angular velocity ω. Then:

(1) the points M_1, M_2 and M_3, symmetric to the point M relative to the straight lines BC, CA and AB move around the circle (with angular velocity $-\omega$);

(2) these three circles intersect at a single point H, the orthocenter of the triangle ABC (**3.8**(b));

(3) each straight line M_iM ($i = 1$, 2 or 3) rotates with angular velocity $(-\omega/2)$ about H;

(4) three points M_1, M_2, M_3 lie on a single straight line l_M passing through H (i.e., the three straight lines M_iM are in fact a single line l_M);

(5) the midpoints of the segments MM_i ($i = 1, 2, 3$) and the midpoint K of the segment MH lie on a single straight line, the Wallace-Simson line;

(6) the point K moves around the circle γ similar to the circle described with magnification ratio 1/2 and center of similitude H;

(7) the circle γ passes through the nine points mentioned in part (b) of Problem **7.23**;

(8) the envelope of the straight lines l_M is a Steiner deltoid that is tangent to the circle γ.

APPENDIX A

Summary of Results from Analytic Geometry

The choice of a coordinate system Oxy in the plane defines an ordered pair of numbers corresponding to each point in the plane; the ordered pair gives the coordinates of the point. The correspondence between the points of the plane and the pairs of numbers is one-to-one: to each point in the plane there corresponds a unique pair of numbers and vice-versa.

1. The distance between the point $A(x_1, y_1)$ and $B(x_2, y_2)$ is determined by the formula

$$AB = \sqrt{(x_1 - x_2)^2 + (y_1 - y_2)^2}.$$

2. The set of points (x, y) whose coordinates satisfy the equation $(x-a)^2 + (y-b)^2 = r^2$ (where a, b, and r are given numbers and $r > 0$) is a circle of radius r with its center at the point (a, b). In particular, $x^2 + y^2 = r^2$ is the equation of a circle of radius r with its center at the origin.

3. The midpoint of the segment between the points $A(x_1, y_1)$ and $B(x_2, y_2)$ has the coordinates $\frac{x_1+x_2}{2}, \frac{y_1+y_2}{2}$. In general, the point dividing the segment AB in the ratio $p : q$ (where p and q are given positive numbers) has the coordinates $\frac{qx_1+px_2}{q+p}, \frac{qy_1+py_2}{q+p}$. These formulas assume a particularly simple form if p and q are selected so that $q + p = 1$.

4. The set of points whose coordinates satisfy the equation $ax + by + c = 0$ (where a, b, c are numbers, and where a and b do not vanish simultaneously, i.e., $a^2 + b^2 \neq 0$) is a straight line. Conversely, each straight line may be defined by an equation of the form $ax + by + c = 0$. In this case the numbers a, b, and c are determined for the given straight line uniquely, apart from a constant of proportionality: if they are all multiplied by the same number k ($k \neq 0$), then the equation $kax + kby + kc = 0$ thus obtained also determines the same straight line.

The straight line divides the plane into two half planes: the set of points (x, y) for which $ax + by + c > 0$, and the set of points (x, y) for which $ax + by + c < 0$.

5. The distance $\rho(l, M)$ of the point $M(x_o, y_o)$ from the straight line l, given by the equation $ax + by + c = 0$, is given by the formula

$$\rho(M, l) = \frac{|ax_o + by_o + c|}{\sqrt{a^2 + b^2}}.$$

This formula assumes a particularly simple form if $a^2 + b^2 = 1$.

Any equation $\alpha x + \beta y + \gamma = 0$ $(\alpha^2 + \beta^2 \neq 0)$ of a straight line may be reduced to this particular form, by multiplying it by either of the numbers

$$\frac{1}{\sqrt{\alpha^2 + \beta^2}} \quad \text{or} \quad -\frac{1}{\sqrt{\alpha^2 + \beta^2}}.$$

Some Facts from School Geometry

B.1 Proportional segments

1. *A theorem on proportional segments.* Let l_1 and l_2 be two straight lines; suppose several segments are marked off on l_1 and parallel straight lines intersecting l_2 are drawn through the endpoints of those segments. The parallel lines then cut off on l_2 segments proportional to the segments marked off on l_1.

2. A straight line parallel to one side of a triangle and intersecting its other two sides cuts off from the triangle a triangle similar to it.

3. *A theorem on the angle bisector in a triangle.* In a triangle, the bisector of any one angle divides the opposite side into segments which have the same ratio as the sides adjacent to the angle.

4. *A theorem on proportional segments in a circle.* If two chords AB and CD of a circle intersect at the point E, then

$$|AE| \cdot |BE| = |DE| \cdot |CE|.$$

5. *A theorem on a tangent and a secant.* If through a point A outside a circle a tangent AT and a secant cutting the circle at the points B and C are drawn, then

$$|AT|^2 = |AC| \cdot |BC|.$$

Notes

1. The theorem on proportional segments is reformulated in the language of motion (pp. 11–12) as the "theorem about the ring on a straight line." A more general assertion, deduced from the theorem about the ring, is given in the lemma on p. 31.

3. The theorem on the bisector of an angle in a triangle has been proved in Problem **2.5** (p. 38) in a more general form for the "cross bisector" which is defined in Proposition **B** of our alphabet (p. 20).

5. The theorem on a tangent and a secant is not referred to anywhere in the book directly but it is closely related to the problems on the radical axis (p. 24).

B.2 Distances, perpendiculars

1. Given a straight line l and a point A not on l, consider perpendicular to l passing through A. The distance from A to the foot of the perpendicular is less than the distance from the point A to any other point on l.

2. A tangent to a circle is perpendicular to the radius drawn to its point of contact.

3. Of two line segments drawn from a given point to a given straight line l, the one which has the larger projection on the straight line l is the longer.

4. (a) If a point lies on the perpendicular bisector of a segment, then it is equidistant from the endpoints of the segment.

(b) If a point is equidistant from the endpoints of a segment, then it lies on the perpendicular bisector of the segment.

These two theorems may be combined in a single statement: the set of all points equidistant from the endpoints of a segment is the perpendicular bisector of the segment.

5. (a) If a point lies on the bisector of an angle, then it is equidistant from the sides of the angle.

(b) If a point included in an angle (smaller than a straight angle) is equidistant from the sides of the angle, then it lies on the bisector of the angle.

From (a) and (b) it follows that: the set of all points contained in an angle (smaller than a straight angle) equidistant from the sides of the angle is the bisector of the angle.

6. One and only one circle can be inscribed in a triangle. This circle is called the *incircle*.

7. One and only one circle can be circumscribed about a triangle. This circle is called the *circumcircle*.

Notes

1–2. These statements may serve as simple illustrations of the tangency principle formulated in Chapter 5 (see the Section on the Extrema of Functions): Suppose straight line γ and a point A is given. Construct the level curves of the function $f(M) = |AM|$; they will form a family of concentric circles. The point on γ at which the function f attains its minimum value is the point of tangency of one of the circles of our family with the straight line γ.

3–4. The general statement of 4 is Proposition A (p. 19) of the "alphabet." The perpendicular bisector is often called the midperpendicular. Statement 3 is essentially contained in the statement of Proposition A on the division of the plane into half planes.

5. A more general statement is formulated in Proposition B of the "alphabet," where the term "cross bisector" is introduced (p. 36).

6. The center of an inscribed circle is determined in Problem 3.3 (p. 37).

7. The center of a circumscribed circle is determined in Problem 3.1 (p. 35).

B.3 The circle

1. Given a circle and a chord passing through two points on the circle, the radial line perpendicular to the chord bisects the chord.

2. *A theorem on tangents.* Fix a point A and a circle γ. Suppose that two tangents AT_1 and AT_2 are drawn to the circle, where T_1 and T_2 are the points of contact. Then $|AT_1| = |AT_2|$.

3. *A theorem on the circumscribed quadrilateral.* A circle can be inscribed in a convex quadrilateral if and only if the sum of the lengths of two opposite sides of the quadrilateral is equal to the sum of the lengths of the other two opposite sides.

4. The set of all the vertices of a right triangle with a given hypotenuse AB is a circle of diameter AB (with the points A and B excluded).

5. *A theorem on an inscribed angle.* In degree measure, the magnitude of an inscribed angle is equal to half the magnitude of the intercepted arc. (In other words, the magnitude of an inscribed angle is half the measure of the central angle intercepting the same arc.)

6. The degree measure of the angle formed by a tangent line and a chord through the point of tangency is half the degree measure of the arc intercepted by this angle.

7. An angle with its vertex inside a circle determines two different arcs: one enclosed between the sides of the angle and the other between the extensions of each side. The degree measure of the angle is half that of the degree measure of the sum of those two arcs.

The degree measure of the angle formed by two secants intersecting outside a circle is half the degree measure of the difference between the intercepted arcs contained by the angle.

8. *A theorem on the inscribed quadrilateral.* A circle can be circumscribed about a quadrilateral if and only if the sum of two of its opposite angles (in degrees) is equal to $180°$.

Notes

4. This statement is discussed on p. 12 in connection with the problem about the cat.

5. The theorem on the inscribed angle is reformulated in the language of motion as the "theorem about a tiny ring on a circle." A more general statement deduced from the theorem about the ring is given in Proposition $\mathbf{E}°$ of the "alphabet."

6–7. Problem **2.6** touches on these theorems.

B.4 Triangles

1. *A theorem on the exterior angle.* An exterior angle of a triangle is equal to the sum of the nonadjacent interior angles.

2. *A theorem on the medians.* The three medians of a triangle intersect at a point which divides each of them in the ratio $2 : 1$ (measured from the vertex). Their point of intersection is often called the *centroid* or *center of gravity* of the triangle.

3. *A theorem on the altitudes of a triangle.* The three altitudes of a triangle are concurrent.

4. *Pythagorean Theorem.* The square of the hypotenuse of a right triangle is equal to the sum of the squares of the legs.

5. The legs of a triangle are proportional to the sines of the opposite angles.

6. The area of a triangle is equal to one-half of the product of:
(a) the base and the altitude intersecting that base;
(b) two sides and the sine of the angle between them.

Notes

2–3. The proofs of these theorems are given on pp. 61–65 in the solutions of problems **3.2** and **3.4** (the fact that any one median divides another median in the ratio 2 : 1 may be obtained from the solution of **3.4**).

APPENDIX C

A Dozen Assignments

This appendix is intended for readers who, after first going through the book and trying to solve the problems that appealed to them, found a number of the problems particularly challenging, and now, in order to understand some of those subtler ideas, are ready to work through the book systematically with pencil and paper in hand.

The twelve assignments given below cover different facets of the book. More specifically, they stress the hidden relationships among the various problems given in each chapter.

In their construction, the assigments follow the standard model of the mathematical correspondence school of Moscow State University. First, the subject matter of an assignment is explained and pages of the text containing the relevant theorems or exercises are given. Then there is a series of exercises, in which the fundamental problems are differentiated from the supplementary ones by the sign ‖. Some of the problems are accompanied by hints or explanations. As for the solutions, we advise you to try to write them out concisely, without unnecessary details, clearly stating the basic steps involved and any references to theorems from your geometry course. Do not forget about particular cases: sometimes they have to be analyzed separately (as in Problem **1.1**, when the point M lies on the straight line AC, or in Problem **1.3**, in the case of a square). Although we are not suggesting that readers give superfluous details when investigating and rigorously analyzing all special cases, we do advise them to give a precise and complete formulation of the result, as is the custom among mathematicians.

C.1 Name the "letters"

The aim of this exercise is to make a first acquaintance with our "alphabet," i.e., with the theorems dealing with sets of points which are useful in the solution of the later problems.

Go through Chapter 2 and make a list of the propositions from **A** to **J** of the "alphabet" on a separate sheet of paper. Against each letter write down the formula (see pp. 33–34) and draw the corresponding diagram.

2.1, 2.2, 2.3, 2.4, 1.16 (a), (b), **5.4** (a), **1.11, 1.12** ‖ **2.13,** | **2.15, 2.16, 3.6.**

Remarks

In the first five problems, we need only state the appropriate letter of the "alphabet" in our answer.

141

Problem **1.16** (a) helps one to solve Problem **5.4** (a) without any calculations.

In construction problems, everything reduces to the construction of a certain point—the center of a circle, etc. The required point is obtained by intersecting two sets from the "alphabet" (see **1.4**). It is essential to name these sets (propositions of the "alphabet") and indicate how many answers the given problem has.

A short solution to **2.13** is based on the result of **2.12**.

C.2 Transformations and constructions

The problems in this assignments involve the various geometric transformations of the circle and the straight line that are discussed on pp. 13–14; they also appear often throughout the book (**6.9** (a), (b), **7.1** (a), (b)).

1.20, 1.21, 1.22, 1.23, 1.24 (a), (b) ‖ **3.7** (a), **4.8** (a).

Remarks

1.22. See the solution to Problem **1.7** (a).

1.23. See the solution to Problem **1.6**.

1.24 (a). Give the answer only.

3.7 (a). Use the fact that the centroid divides the median in the ratio 2 : 1, as measured from the vertex.

4.8. Read the solution to Problem **4.7**.

In all these exercises we suggest that you make sketches of all necessary constructions. Write your solutions concisely, paying attention to the sets and transformations used. Indicate how many solutions each problem has.

C.3 Rotating straight lines

This assignment primarily concerns the different variants of the theorem on the inscribed angle and its corollaries.

Go through the book in the following order: Problem **0.1** (about the cat), Problem **1.1**, the theorem about the ring on a circle (pp. 11–12), Propositions **E°** and **E** of the "alphabet" (pp. 21–22). Note that the theorem about the ring (and the problem about the cat) should not be interpreted literally: the imagined "ring" is simply the point of intersection of the straight line and the circle; if we made a wire model, then after a single rotation (in either direction) the ring would get stuck and stop moving.

1.8, 1.9, 1.10, 1.13, 1.18, 2.6 (a), (b) ‖ **1.27, 2.7, 2.8** (b), **4.6, 7.5, 7.6**.

Remarks

1.9. Draw diagrams for the different positions of the point A.

1.10. Draw a straight line through the point B, plot the point A' symmetric to the point A with respect to this straight line, and then draw the segment BA'.

Show the sets of points in the answers to problems **1.8** and **1.10** in the same diagram. By what transformation can one get the set in **1.10** from the one in **1.8**?

1.13. State how many answers the problem has.

1.27. Carry out an experiment using an ordinary T-square. Hint: circumscribe a circle about the wooden triangle, join the vertices of the right angles, and use the theorem on the inscribed angle.

2.6. Imagine that the movable chord is moving uniformly around the circle.

2.8. (b) The solution is analogous to **2.8** (a). Look at the second variant of the solution of this problem, given on p. 42.

C.4 Straight lines and linear relations

The problems in this assignment deal only with straight lines.

Go through the book in the following order: problems **1.2** and **1.3** about the bicycle and the rectangle (pp. 8–11), the theorem about the ring on a straight line (pp. 11–12), and the important lemma (p. 31) which extends it, and also Propositions **F, I, J** of the "alphabet" and the general theorems on the distances to straight lines and the squares of distances (pp. 23–33).

1.24 (a), (b), **2.18**, **2.19** (b), **3.9**, **3.14**, **3.15**, **3.16** ‖ **1.26**, **1.27**, **2.14**, **2.20** (a), **3.18**.

Remarks

2.18. See solutions to **2.5** and **2.17**.

2.19 (b). Find out how the answer depends on the dimensions of the rectangle $a \times b$ and the parameter μ (see the answer to Problem **2.19** (a)).

3.14–3.16. See Proposition **C** of the "alphabet."

1.27. Let a and b be the lengths of the legs of the wooden triangle. Find the ratio of the distances from its free vertex to the sides of the given right angle.

2.20. It is sufficient to give the answer and a diagram.

3.18. Read the solution to **3.17**.

C.5 The tangency principle (conditional extremum)

The assignment consists of problems on finding maxima and minima. Every problem may be reduced to that of finding the particular point on some line or curve (as a rule, one of the sets from the "alphabet") at which a given function reaches its maximum or minimum value. Read the solution to problems **4.1, 4.2** and **4.7** (pp. 47–49), the solution to Problem **5.1**, and the rest in Chapter 5, particularly pp. 64–65. Study (or redraw) the maps of the level curves on pp. 60–61.

Remarks

5.4 (a). See Problem **1.16** (a).

C.6 Partitions

In this assignment we find various sets of points satisfying inequality constraints and we also specify the set operations (intersection or union) that correspond to the logical combinations of the different constraints. Many propositions of our "alphabet" in Chapter 2 have conditions of the following type: the curve consisting of the points M for which $f(M) = a$ divides the plane into two domains, one in which $f(M) < a$ and the other in which $f(M) > a$ (here f is some function on the plane; see p. 57). In exactly the same way, if f and g are two functions on a plane, then the set of points M, where $f(M) = g(M)$ partitions the plane into regions. In some of these, $f(M) > g(M)$, while in others, $f(M) < g(M)$. Go through the text in Chapter 3 (pp. 39–40), and the solutions to problems **3.11, 3.23** (about the cheese).

   1.19, 3.12, 3.14, 5.3 (a), (b), **3.15, 3.16 ‖ 3.18, 3.19, 4.11, 4.12** (a), (b).

Remarks

   1.19. Draw the segment BC and indicate the set of vertices A of the triangles ABC for which each of the conditions (a), (b), (c) is fulfilled. Use the second paragraphs of Propositions **D** and **E** of the "alphabet."

   3.14. Read the solution to **3.13.**

   3.15–3.16. For each side of the polygon, construct the strip as in Proposition **C** corresponding to $h = S/p$. Can these sets cover the entire polygon whose area is S?

   4.11–4.12. Read the solution to **4.10.**

C.7 Ellipses, hyperbolas, and parabolas

The aim of this assignment is to acquaint ourselves with the first definitions of these curves, given in Propositions **K, L, M** of our "alphabet." Go through Chapter 6 and list the propositions of the "alphabet." For each letter, write down the formula and draw the corresponding diagram (problems **6.5** (a), (b) of this exercise will help you do this).

   6.1 (a), (b), (c), **6.2, 6.3** (a), (b), (c), (d), **6.4** (a), (b), **6.5** (a), (b), **6.10** (a), (b), **6.11** (a), (b) ‖ **6.8, 6.12** (a), (b), **6.13** (a), **6.14, 6.24.**

Remarks

   6.1 (a), (b), (c). Indicate how the answer depends on the parameter (put $|AB| = 2c$).

   6.2. Use the theorem on the segments of the tangents to a circle.

   6.4 (b). Consider positions of the quadrilateral $ABCD$ for which the link BC crosses AD.
   The following problems deal with the focal properties of the curves.

   6.10 (a). The proof here is similar to the one in the solution of **6.9** (a), and is also based upon Problem **6.7.**

6.11 (a). Compare the definition of a parabola (Proposition **M** of the "alphabet") and its focal property.

6.8. The proof is similar to the proof of the orthogonality of equifocal ellipses and hyperbolas (pp. 71–72).

C.8 Envelopes, infinite unions

The problems in this assigment are all fairly complicated. Each problem concerns an entire family of straight lines or circles. If we take the union of the lines or curves in such a family, we obtain an entire region of the plane. It often happens that the boundary of this region is the envelope of the corresponding family of lines— that is, a curve (or a straight line) which is tangent to all the lines in the family. (For example, in the solution to Problem **1.5** on p. 13, we used the fact that the envelope of the family of chords of equal length of the given circle is a circle concentric to the given circle.) We urge you to draw a diagram for each problem. It is not necessary, however, to draw the envelopes. If you draw a large enough number of lines of the family, then the envelopes appear more or less "automatically" (as in the diagrams on pp. 76–77).

Read the text on pp. 121 and 15, the solutions to **3.20** (b), **6.6**, **6.7** and the proof of the focal property of a parabola (pp. 72–73).

1.30, 3.20 (a), **3.22, 4.5, 6.16** (a), (b), **6.17** ∥ **6.15** (a), **6.25** (a), (b), **7.2, 7.20.**

Remarks

3.20. Imagine this union as the set of vertices M of a hinged parallelogram $OPMQ$ with sides 3 cm and 5 cm; compare this method with Chapter 7 (pp. 92–93).

3.22. If, in the first t minutes, the man walks along the road, and then for the next $60 - t$ minutes he walks through the meadow, where will he end up? Now take the union of the sets obtained for all t from 0 to 60.

4.5. What kind of set appears the answer to Problem **3.22** if we replace one hour by T hours? Find for the value of T for which this set contains the point B.

7.20. The family of tangents to the nephroid has been considered in Problem **7.16**. Also recall problems **7.1** (a), **7.2** on the cardioid, and the theorem about two circles (pp. 99–101).

C.9 Tangents to cycloids

This assignment includes a series of problems in which one has to prove that the envelope of some family of straight lines is a cycloid. The solutions to most of these problems are based on the theorem about two circles. Read the statement, examples, and applications of this theorem on pp. 99–100. Also, carefully analyze its proof on pp. 104–105.

7.17 (a), (b), **7.16, 7.18, 7.19** ∥ **7.21, 7.22, 7.23.**

Remarks

7.17. Find the curve along which the endpoints of the diameters move and find the curve that is their envelope. (Compare the result with the last diagram on p. 97.)

7.16. Using the theorem about two circles, describe the family of tangents to the nephroid. Find the solution to Problem **7.15.**

C.10 Equations of curves

The method of coordinates allows us to make very natural generalizations of particular geometric observations (go through the general theorem in Chapter 2, pp. 25–26 and 31–32, Chapter 6, pp. 78–88). The representation of curves in equation form allows us to solve geometric problems with the language of algebra. In this assignment, there are exercises on the method of coordinates and problems in which it is used in a natural way. Most of the problems are related to second-order curves. In some problems it is necessary to change from parametric equations to algebraic ones (see the solution to **0.2**, p. 3).

1.16 (c), **6.18**, **6.19** (a), (b), (c), **6.20** (a), (b), **7.24** (b) ‖ **6.21** (a), (b), **6.23**, **6.25** (a), **6.26** (a), (b), **6.27.**

Remarks

In the problems about distances to points and straight lines, you must investigate carefully how the answer depends on the parameter. For each of these problems, you must draw the corresponding diagram—the family of curves. It is convenient to draw the ellipse according to the given equation, representing it as a compressed circle (p. 79) and the hyperbola by drawing its asymptotes and marking its vertices (the points of the hyperbola closest to its center).

In Problem **6.26**, if we limit ourselves to points M lying inside the triangle, then a beautiful geometric solution may be given using similar triangles, as well as the theorems on the inscribed angle and the angle between a tangent and a chord.

C.11 Geometrical practical work

In this assignment you have to construct diagrams which illustrate the most interesting definitions and properties of curves; hopefully, this will give you a vantage point for the whole book!

It is said that "Geometry is the art of reasoning correctly given an incorrect diagram." Often, however, it is valuable to have the same approach to geometry as to physics: an exact diagram is a geometrical experiment. Such an empirical view helps us analyze difficult statements about families of lines or complicated configurations, and, in turn, to discover some new regularity.

We advise you to redraw (sometimes with elaborations) those diagrams which depict interesting families of straight lines and circles. Technically speaking, such illustrations are relatively simple to create, but nevertheless, accuracy and a certain ingenuity are required to make them beautiful. On large sheets of paper, your drawings will look considerably more significant than our little diagrams in the margins.

1. *The astroid* (p. 5). Try to make sure the midpoints of the segments are uniformly distributed around the circle on which they lie. The larger the number of segments drawn, the more obvious their envelope, the astroid, will be.

2. *Orthogonal families of circles* (p. 59). The first family is the family of all possible circles passing through the points A and B (see **2.1**). The second family is the family of circles whose centers lie on the straight line AB. If M is the center of one such circle, then the circle's radius is the segment of the tangent drawn from the point M to the circle with diameter AB.

3. *Ellipses, hyperbolas and parabolas* (p. 70). The method of construction is mentioned in problems **6.5** (a), (b). Color the "squares" alternately with two different colors, as on a chessboard (see p. 16 and the remarks on Problem **6.8**). Make another copy of each of the diagrams to problems **6.5** (a), (b), and, using ink, mark on them the families of ellipses, hyperbolas and parabolas.

4. *Second-order curves as the envelopes of straight lines* (pp. 76–77), Figs. 4–6). The method of construction follows from **6.16** and **6.17**.

5. *Rotating straight lines.* Make your own diagram, illustrating Proposition $E°$ of the "alphabet" (the lower diagram on p. 21). Draw a circle and divide it into 12 equal parts. Draw straight lines through one of the points of division A and the other points of division and also the tangent to the circle at the point A. The result is a bundle of 12 straight lines dividing the plane into 24 angles of equal size). By moving a pencil around the circle, we can see that whenever we go from one point of division M to the next, the straight line AM rotates through the same angle. Choose another point of division B (say, the fourth point from A) and a corresponding bundle of 12 straight lines similar to the one for the point A. For each point of division M, mark out the acute angle between the straight lines AM and BM. (All these angles are equal!)

From Theorem $E°$, it follows that if all 23 straight lines constructed are extended to their points of intersection, then all 110 points of intersection thus obtained (not counting the points A and B) lie on 11 circles, 10 on each circle ⟨?⟩.

Using two different colors (white or black) alternately fill in, as on a chessboard, the "squares" of the net you obtain. You will then immediately see the family of circles passing through the points A and B and the family of hyperbolas (it is even transparent if you take a bundle of 24 straight lines). For, if straight lines passing through the points A and B rotate in opposite directions with equal angular velocities, then their point of intersection will move along a hyperbola (**6.23**).

6. *The conchoid of Nicomedes and the limaçon of Pascal* (pp. 87–88 and 94). The conchoid of Nicomedes is obtained in the following manner. A straight line L and a point A are given. On every straight line l passing through A, mark off two segments, each of constant length d—that is, one segment of length d in each direction—from the point of intersection of l with L. For each different value of d, draw the family of such conchoids.

The limaçon of Pascal is obtained in a similar manner. Suppose we are given a circle γ and a point A on it. On every straight line l passing through A, mark off two segments, each of constant length—again, one segment in each direction—from the point of intersection of l with γ of constant length.

7. *The cardioid and the nephroid as the envelopes of the circles* (p. 91, **7.2** and p. 105, **7.20**).

8. *The cardioid and the nephroid as the envelopes of reflected rays* (the drawings on pp. 100 and 101). It is convenient to construct these drawings using the fact that the chord of the incident ray is equal in length to the chord of the reflected ray.

C.12 Small investigations

Almost any problem in geometry is a subject for independent research, demanding both inventiveness and originality of thought. In this assignment, we highlight four difficult problems whose solutions require the use of a wide range of different arguments and techniques.

4.12 (a), (b), **4.14** (a), (b), **6.15** (a), (b), **7.23** (a), (b).

The solution to Problem **4.14** (b) is very similar to the solution for the problem about the motorboat.

Index

About Victor Gutenmacher

Victor Gutenmacher is a distinguished mathematician and educator with extensive research and teaching experience in algebraic topology, geometry, and numerical methods. He received his Ph.D. in mathematics from Rostov and Moscow Universities in 1974, and he was a Senior Researcher and Professor of Mathematics at Moscow University from 1969 to 1988.

Dr. Gutenmacher has expertise in applied mathematics, software engineering, and computer-aided design. He has conducted research in topology, geometric modeling, and mathematical economics. He has over twenty years of teaching experience at all levels, from secondary school to graduate school. He has taught undergraduate and graduate courses at Moscow University in a wide range of subjects, such as abstract algebra, calculus, discrete mathematics, single- and several-variable complex analysis, mathematical programming, and mathematical methods in economics. In addition to *Lines and Curves*, Dr. Gutenmacher, N. B. Vasilyev, and their colleagues J. M. Rabbot and A. L. Toom wrote the Russian text *Mathematical Olympiads by Correspondence*. (In fact, many of Dr. Gutenmacher's papers with N. B. Vasilyev were written under their collective pen name "Vaguten," which blends both their last names.) Dr. Gutenmacher also coauthored *Homotopic Topology* with A. T. Fomenko and D. B. Fuchs.

For the past 15 years, Dr. Gutenmacher has worked in the United States as a mathematician and senior software engineer at Computervision, Auto-Trol Corporation, Structural Dynamics Research Corporation, and, currently, at VISTAGY, Inc. He is presently involved in the development of a computer aided design (CAD) system-independent geometry engine; this geometry engine enables VISTAGY's world-class products to be tightly integrated with all of the major high-end CAD systems. Furthermore, for several years, he was also Senior Mathematics Consultant at BBN Technologies in Cambridge, Massachusetts.

Dr. Gutenmacher is the author of more than 80 publications in mathematics and mathematics education. He was a member of the Advisory Panel on the Committee for American Mathematics Competitions. In Russia, he served as a member of the Editorial Board of *Quantum* from 1981 to 1988; the chairman of the Methodology Committee for the Gelfand Correspondence School from 1969 to 1988; a member of the Methodology Committee for the USSR Mathematical Olympics from 1966 to 1979; and the coach for the Soviet team in the International Mathematical Olympiad from 1973 to 1979. He is a member of the American Mathematical Society.

About N. B. Vasilyev

On May 28th, 1998, N. B. Vasilyev, one of the two authors of *Lines and Curves*, died at the age of 57 after a long and debilitating illness. He was an extraordinary intellectual: a talented mathematician, encyclopedically educated scientist, and renowned educator.

He graduated with honors from the Moscow Conservatory School of Music. However, he chose to work in mathematics rather than music; in 1957 he enrolled as a freshman in the Mechanics and Mathematics Department of Moscow University, and he graduated in 1962. Until his untimely death nearly forty years later, he remained closely associated to the University.

After completing his postgraduate studies, he began a successful career in the Department of Mathematical Methods in Biology at the A. N. Belozersky Institute of Physico-Chemical Biology, Moscow State University, where he joined a promising group of young scientists headed by I. M. Gelfand. Vasilyev started his record of mathematical publications at MSU, and continued to publish prolifically throughout his life.

His other lifelong commitment—the mathematics education of schoolchildren—began during his freshman year at Moscow University, when he became a member of the Organizational Committee for the Moscow Mathematical Olympics. As a committee member, he wrote and collaborated on the creation and grading of examinations and student submissions.

At that time there was great interest in mathematics and physics in the Soviet Union, especially among schoolchildren. Faculty members at Moscow University played an active role in the mentoring of a new generation of future scientists, particularly through the National Mathematical Olympics and the Mathematical Club for Schoolchildren. Vasilyev became one of the leaders of and key contributors to both organizations; for over ten years, he was the vice president of the National Mathematical Olympics Committee, chaired by A. N. Kolmogorov.

Vasilyev also served on the examination panel of the Moscow University Mathematics and Physics Preparatory School for many years. Along with I. M. Gelfand and I. G. Petrovsky, he helped found the National Mathematical Correspondence School. He was one of the creators of *Quantum*, a national mathematics journal for young adults, and served as the publishing director of its most demanding department, "Quantum Tests." It would not be an exaggeration to say that every major development in mathematical education in Russian between the 1960s and 1990s had N. B. Vasilyev's expert involvement and tireless participation.

Special notice should be given to Vasilyev's talent for, dedication to, and lengthy career in the popularizing of mathematics. He had an unsurpassed ability to define concepts and mathematical problems concisely and beautifully; both his writing and his lectures exemplified simplicity without oversimplification and depth without excess. His articles and presentations were models of clarity and artistry.

The book *Lines and Curves* grew out of a collection of assignments written by Nikolay Vasilyev and Victor Gutenmacher for the National Mathematical Correspondence School. Both authors were instrumental in the founding and growth of the mathematical division of the Correspondence School from its very inception in 1964. *Lines and Curves* was eventually published and subsequently edited for a second publication as part of the series "The Library of the School of Mathematics and Physics" under the direction of I. M. Gelfand.

This book is the result of painstaking and inspired work by two friends and colleagues. It is a masterful accomplishment, the transformation of what was originally an assortment of geometry problems into a fascinating text to which many generations have paid tribute. I was a witness to the creative collaboration that brought about *Lines and Curves*—a book I very much enjoyed and which, I hope, future readers will as well.

J. M. Rabbot
Moscow, 2000